The Canons of The Sinister Universe:

The Last Four Books on the Universal Model of Our World

Greg Feild

November 5, 2017

Abstract:

This book is a compilation of my final four papers concerning The Sinister Universe.

On Wave Particle Duality and the Quantum of Action

Greg Feild

July 6, 2017

About the author:

I earned a Ph.D in experimental high energy physics from the Pennsylvania State University working on HERA at DESY in Hamburg, Germany studying photoproduction and deep inelastic scattering in electron-proton collisions.

I did my postdoctoral studies with Yale University working at Fermilab on the CDF experiment at the Tevatron. My primary research interest was particle hadronization in quarkonium production in proton-antiproton collisions.

"There's no use trying, one *can't* believe impossible things."

-- Alice to the White Queen

Abstract:

In this paper, we continue our interpretation
of the investigations of physics
as the science of mass in motion.

The universe is sinister,
but it is no longer spooky.

it's all about the oomph

The **NULL** hypothesis:

All physical interactions can be described by

Newton's **U**niversal **L**aws + a **L**agrangian

let's do physics!

Preface:

Whilst googling myself this afternoon (a bad habit I've developed recently), I came across a little theory of gravity called Gravitoelectromagnetism (GEM).

I was like,

"This theory is brilliant!"

:)

"Genius!", I said.

The only thing the GEM equations (analogs of the Maxwell equations) are missing is the mass current term, J_m, in the equation for electromagnetic induction, which we introduced in "On Parity and Isospin".

Then I learned about Gravity Probe B.

Very exciting !

Particle in a box:

```
                                                    ^
                                                     \
                                                      \
          |------------------<-----------------------\|
                                                       ^
                                                        \
                                                         \
          |------------------<------------------------\---|


          |------------------<--------------------/--------|
                                                 /
                                                /
                                               /

node      |----------<----------------.------------------|

                                        ^
                                       /
                                      /
          |----------<-------/-----------------------------|

                                  ^
                                 /
                                /
          |-----<---/--------------------------------------|




          |--------\------------------------>-----------|
                    \
                     \
                      \

artist's conception
```

Introduction:

In our last book, "A Critical Examination of Classical and Quantum Mechanical Waves", we proposed a mechanical model for the wavelike nature of the electron.

In our new model, the electron is an inertial (i.e. spin ½) blob of mass-energy/charge, spinning to the left. The axis of the electron spin and the principal axis of rotational inertia of the electron are aligned or 'projected' along the direction of motion of the electron and precess about this direction with a frequency proportional to the total mass-energy of the electron.

This precession frequency corresponds to the de Broglie wavelength of the electron;

$$v = E/h == m/h \qquad (1)$$

$$\lambda = h/p = h/mv \qquad (2)$$

with wavenumber

$$k = 2\pi/\lambda \qquad (3)$$

and so it takes 2π radians for electron spin to precess once about the direction of motion.

However, since the electron is a 'spinor', it takes 4π radians, or two revolutions of the spin vector, for the precessing *angular momentum vector* to return to its original value and helicity.

A free electron of fixed helicity, executes a 'polarization flip' every 2π radians, performing a 'complete revolution' every 4π radians.

This spin flipping is what gives a particle its oomph!

A free photon also performs a polarization flip every 2π radians.

That's the theory at least.

In addition, in our model, an electron's resistance to acceleration is a consequence of the conservation of angular momentum.

It's hard to shift about things that are spinning, and electrons are things that are spinning!

Particle in a box:

In "A Critical Examination of Classical and Quantum Mechanical Waves", we considered the classic example of the particle in a one dimensional box in light of our new theory.

A particle (e.g. an electron) in a box bounces back and forth between the walls. The particle follows a well defined path and physically traverses *all points* lying between the walls of the box.

Figure 1 shows the probability distribution for the state n = 2.

As we demonstrated in "Critical", the electron spin precesses around the direction of motion, performing a helicity flip every 2 pi radians, and completing one complete 'revolution' every 4 pi radians.

At the point x = L/2, the electron is undergoing a flip in helicity, where its spin is effectively zero, and hence it cannot interact with our experimental probe (e.g. a photon).

Similarly, at the walls of the box, the electron must perform a helicity flip as it bounces, exchanging a virtual photon with an electron in the wall of the box.

At the point x = L/4, the electron helicity is in 'full bloom' and it can interact with a real photon.

The electron oscillates harmonically between the ability to engage in real and 'virtual' interactions

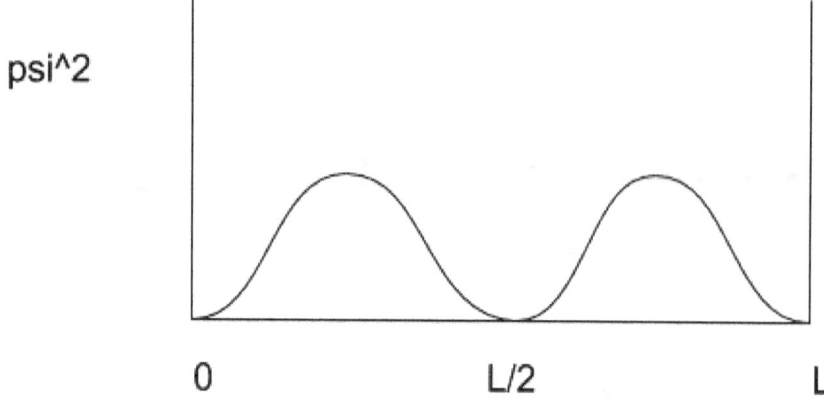

Figure 1: Probability distribution for a particle in a box; n = 2.

Quantization:

Ultimately, quantization boils down to the satisfaction of boundary conditions.

Virtual photons can only attach to/interact with an electron when the electron has an effective helicity of zero (during its polarization flip) because virtual photons are spin zero.

This fixes the 'frequency' of the virtual photon coupling two interacting electrons, since virtual interactions, or the exchange of energy and momentum, can only occur when both electrons have helicity ~= 0.

When an electron reaches 'peak helicity', it is able to interact with (i.e. absorb and emit) real photons the most readily and the most efficiently.

Cosmic springs:

This view of quantization bolsters our conclusion that all interacting particles are constantly coupled by 'dedicated' virtual photons. These photons would vary in mass and frequency during an interaction in such a manner as to constantly satisfy the uncertainty principle.

Virtual photons *do not* radiate from a particle in a manner analogous to Faraday's lines of force. This radial representation is only valid for classical fields OR a source of real photons such as a radio antenna.

We argued in "Observations on the Quantum Mechanical Nature of Gravity", that particles do not send out 'feeler photons on spec", looking for something to couple to. This would require an infinite amount of time and energy, *and* chances are that most of these virtual photons would find *no target*.

When would a particle 'decide' it is time to give up and reel the virtual photon back in?

Entanglement:

The assumption behind things like entanglement and the EPR paradox is that particles do not have definite physical properties until they are measured!

Why anyone would think such a crazy thing is unclear.

In our theory, particles are real physical objects. They always have a definite position, momentum, and spin orientation.

We can only make probabilistic determinations of these quantities because particles are *very small*.

If one prepares two 'correlated' particles, for example, one with spin up and one with spin down, then one particle will always have spin up and the other spin down.

This will be true always and forever, no matter how far apart the particles may stray.

Quantum entanglement is not a thing.

The integral over all paths:

In the classical formulation of the Hamiltonian and Lagrangian approach to the dynamics of a system, one performs an integration over all possible paths in order to minimize the action.

One laughs and says, "This is crazy, I know the particle does no such thing!" However, the method gives the desired result, so one shrugs and moves on.

In quantum mechanics, of course, the integral over all paths formalism is accorded some mysterious and mystical significance. The particle is actually thought to follow all the possible paths available to it, and all at the same time !!!

We can be somewhat sympathetic to this view, since it was proposed in the bad old days of field theory.

Now it is another quantum boogeyman we must reject.

Hidden Variables:

We have discovered the hidden variables of quantum mechanics.

They are called spin, mass, and charge!

In our model, subatomic particle interactions are completely deterministic. In addition, there exist many physical and 'timing' constraints on these interactions, since the electron, as a spinor, can only interact at certain times, and in certain ways, as its helicity oscillates back and forth as it travels a *well defined* path.

In "Critical", we demonstrated that the Bohr radius of the electron for the hydrogen atom is *real*, *and,* that the electron travels a well defined, circular, and *closed* path as it orbits around the proton.

Reversibility:

In principle, the equations of classical and quantum mechanics are invariant under 'time reversal'; causing one to wonder why time seems to flow in only one direction in our physical universe.

In principle, the equations describing the earth's orbit around the sun are invariant under time reversal and space reflection.

In practice, one would need a really big 'tractor beam', and a whole lot of energy to decelerate the earth and send it off in the opposite direction; all the while trying to keep the earth from spiraling into the sun!

Time flows one way, into the future; unstoppable like a runaway freight train!

The single slit experiment:

In our model, the characteristic 'interference' pattern observed on the screen or photographic plate in a single slit diffraction experiment, is not due to two separate electron waves arriving at the screen out of phase and canceling each other out.

Instead, each separate electron arrives at the screen 'in or out of phase' for being able to interact with the photographic plate.

An individual electron can only darken the photographic plate if it arrives with the proper phase. Just like our particle in a box, the oscillating helicity of the electron has to be in the process of 'flipping polarization' when it arrives at the screen (i.e. the electron must have helicity ~zero to interact with the surface of the plate).

Electrons that do not successfully interact with the surface of the plate will carry on until they *do* find something to interact with.

We suggest a single slit experiment where the photographic plate is replaced with several layers of silicon, analogous to tracking detectors employed in high energy scattering experiments.

Our prediction is, of course, that the electrons which fail to contribute to the pretty pattern on the top layer, will show up in subsequent layers; perhaps displaying secondary pretty patterns.

The double slit experiment

We can employ similar arguments to explain double slit interference patterns, although in reality the situation is much more complicated.

In the single slit experiment, it does not really matter what happens before the slit. The slit is essentially the source of the electrons and the electrons have a characteristic spread determined by the width of the slit and the uncertainty principle.

For the double slit experiment, the environment of the electrons before the two slits *does* matter.

Although an electron can only pass through one slit or the other, it feels the 'potential' of both slits.

Remember, our 'free' electrons are actually exchanging virtual photons with *everything*. On their approach to the two slits, they are interacting with the wall housing the two slits, and they are interacting with the screen (which is their final destination) by virtual photon exchange via the two slits.

How this could actually be cast and analyzed in terms of potentials is unclear!

We suggest a similar double slit experiment, where the the two slits are replaced by two similarly sized and spaced metal plates, set to a positive retarding potential; thus, scattering the incoming electrons back toward the source and an appropriately placed screen.

The quantum of action:

There seems to be no entity called 'the quantum', and no fundamental physical quantity that corresponds to one 'quantum of action' (unless, of course, it turns out to be the electron neutrino!).

The formula relating the energy of a photon to its frequency, $E = h\nu$, shows that, in principle, the photon energy can take on *any* value.

We conclude the Planck constant (although a momentous discovery) is really nothing more than a glorified conversion factor describing the relationship between the frequency of a particle and the energy of a particle.

Any quantization of the energy of a photon, is ultimately due to the quantized nature of transitions in bound leptonic systems which result in photon emission.

The Lorentz force:

The complete, 'classical', relativistic, Lorentz force between two identical electrons is

$$\mathbf{F} = (G/c^4 - (e/m_e)^2(\mu/4\pi))(c/R)^2)(m_1 m_2 \mathbf{r} + (1/c^2)(\mathbf{p1 \times p2 \times r})) \quad (4)$$

where, of course, $\mathbf{F}_1 = -\mathbf{F}_2$.

Newton's first law tells us

$$\mathbf{F}_1 = m_1 \mathbf{a}_1 \quad (5)$$

If we compare equations (4) and (5) we can see that the acceleration of a particle is *independent of its mass*.

$$\mathbf{a}_1 = \text{Function}(m_2, \mathbf{R}) \quad (6)$$

This is a general result that we used to assume applied only to the gravitational interaction.

Now, *that* is mind bending physics!

The bottom line:

   Particle interactions are all about balance.

   Energy and momentum interact through the exchange of energy and momentum; conserving energy and momentum.

   Facetious, but true!

   Why this 'motion'? Why this constant shuffling?
   Why this endless shuttling of energy and momentum
   back and forth; to and fro?

   Is there a method to this madness?

   Is there some equilibrium to be reached?

   Our current model indicates that 'the goal' is a balance between the mass-energy of a particle and the kinetic energy of a particle; or more properly and precisely, the balance between the kinetic and 'potential' energies of a system of particles.

$$\text{goal of universe} == m - m_0 = \text{kinetic energy} = \text{const.} == 42 \qquad (7)$$

            of course!   :)

            The answer is 42.

conclusion:

   just be cool.

   physics is fun!

$$:)$$

gilding the lily: :(

                     physics sells itself

Our troubled times:

When a physical theory, that make *no* physical sense,
is labeled 'mind bending' rather than discarded;

The mind boggles.

If a physical theory bends your mind,
it is wrong;
or at least, incomplete.

Correct theories straighten things out!

It is always better to say, "I don't know", than to spout some gibberish.

Similarly, one should always say "our current model says", rather than
"we now know" (e.g. space is curved, nucleons are made of quarks, etc., etc.);
even now, now that we have a Final Theory of Everything !

This is how reasonable, thoughtful, and *responsible* people use words.

Word.

References:

Modern Elementary Particle Physics
Gordon Kane

Quantum Physics
Rolf G. Winter

Gauge Theories in Particle Physics
I. J. R. Aitchison and A. J. G. Hey

Quarks and Leptons: An Introductory Course in Modern Particle Physics
Francis Halzen, Alan D. Martin

Symmetries and Group Theory in Particle Physics
Giovanni Costa, Gianluigi Fogli

and

Elementary Modern Physics
Richard T. Weidner, Robert L. Sells

a universal feild theory   :)

Books by Greg Feild:

the pentateuch

1. "A quantum mechanical theory of gravitational interactions"
   CreateSpace Independent Publishing, 8/29/2016

2. "Observations on the quantum mechanical nature of gravity"
   CreateSpace Independent Publishing, 10/8/2016

3. "On gravitation and electric charge"
   CreateSpace Independent Publishing, 11/1/2016

4. "On spin, mass, and charge"
   CreateSpace Independent Publishing, 11/29/2016

5. "On angular momentum, acceleration, and absolute motion"
   CreateSpace Independent Publishing, 1/4/2017

the exegeses

6. "The Sinister Universe"
   CreateSpace Independent Publishing, 3/1/2017

7. "On Parity and Isospin"
   CreateSpace Independent Publishing, 4/11/2017

8. "Reflections on the Sinister Universe"
   CreateSpace Independent Publishing, 5/12/2017

the hermeneutics

9. "On Current Physics"
   CreateSpace Independent Publishing, 6/11/2017

10. "A Critical Examination of Classical and Quantum Mechanical Waves"
    CreateSpace Independent Publishing, 6/18/2017

do physics!

Notes:

How I Did It: :)

    Well . . . we certainly did not start from scratch!

    All the pieces of the puzzle had been gathered; they just needed to be put together.

    The realization of the neutrino mass was the tipping point; vaulting us all into the realm of

    the *adjacent possible*.

                                    peace out

On Matter, Mass, and Motion

Greg Feild

September 14, 2017

Thus scientific knowledge is a demonstrative state, . . .

i.e. a person has scientific knowledge when his belief is conditioned in a certain way, and the first principles are known to him; because if they are not better known to him than the conclusion drawn from them he will have knowledge only incidentally.

This may serve as a description of scientific knowledge.

-- Aristotle,
Ethics

About the author:

    Greg Feild is a physicist.

    He has a PhD and everything!

Abstract:

In the universal model of our sinister universe,
nature (i.e. all interaction) is mechanical
and deterministic.

In this book, we explore this strange new world!

In addition, we offer the usual corrections and clarifications
concerning parts of our model still under development,
and summarize the current status of this totally awesome,
completely mind-straightening, universal model of the world.

let's do physics !

G_F

Errata:

To date, we have been ~~insistent~~ positing that the electron neutrino has a magnetic moment, mu_v, and this magnetic moment is proportional to the mass of the neutrino, m_v, the electric charge, e, and an incorrect factor of 1/c for some reason … ;

mu_v = e*hbar/2*m_v*c   (a)

We still expect the neutrino to have a magnetic moment, but the factor of e is not compatible with recent developments in our theory, nor is it 'parallel' with our expression for the gravitational contribution to the generalized Lorentz force; a generalization which was motivated by our assuming a magnetic moment for the spinning mass of the neutrino!

The 'logical' solution is to replace the coupling constant, e, with the square root of the gravitational constant G divided by 4*pi*epsilon_0, and multiply by the relativistic mass (and remove the factor of 1/c).

mu_v = (G/4πε)^½(m)(hbar/2*m_v)   (b)

We will explore how we arrived at this formula later in this paper.

The mass of the electron is still presumed to be

m_e = e*m_v   (c)

(where e is the *magnitude* of the electric charge), but the electron magnetic moment would now be given by

mu_e = e*hbar/2*m_e + (G/4πε)^½(m)(hbar/2*m_e)   (d)

Speaking of the Lorentz force, we've encountered a lot of trouble (made careless errors) in crafting our recommended generalization. Here is the latest offering.

The complete, 'classical', relativistic, Lorentz force between two identical electrons is (now)

F = (G/c² - (e/m_e)²(μ/4π))(c/R)²(m_1m_2**r** + (1/c²)(**p_1**x**p_2**x**r**))   (e)

Finally, what we have been calling the 'matchbook summary', now looks like this;

L_interaction = -i*hbar(G^{1/2} - e/m_rest)(ψ^bar γ^μ A_μ ∂ψ/∂t)   (f)

Now, everything should be parallel, even if not fully correct!

Preface:

Intrinsic angular momentum, magnetic moments, spin, polarization, helicity ....

Confounding and confusing!

And, the foundation of our physical world; the fundamental "expression" of energy and momentum; the very essence of matter.

That's all.

The world is all about spin. Planets, stars, galaxies, gas molecules, electrons, neutrinos, photons; all have 'intrinsic' angular momentum, or spin; a.k.a. *absolute motion*. What things are missing from this list? Balls, buckets, tops, gyroscopes, merry-go-rounds, etc.; all human sized objects. Familiar everyday objects do not spin unless one applies an external torque and, of course, they all will eventually stop rotating due to friction. We hypothesize this is why the importance of the role of spin in both microscopic and cosmological interactions has been generally minimized and overlooked in the past.

I have personally been quite flummoxed by magnetic moments; not only due to an incomplete understanding of the standard model representation, but also in reconciling this representation with our theory that magnetic moments are due to spinning particle mass.

In this current paper, we will reexamine the magnetic moment; salvaging (?) our theory and reconciling it with the current standard model theory, which, of course, is brilliant and correct!

Actually, we will derive the 'relativistic' magnetic moment for the electron, which reduces to the standard model representation as in the first term of equation (d), for an electron at rest.

On the other hand, our notion of the conservation of the "electromagnetic charge" does not seem quite so easy to salvage; or at least not worth the effort at the moment.

(This theory joined the model late in the game, and it really did not contribute very much!)

Our theory of 'weak isospin' should still be strong.

Big-ly!

Introduction:

Where to begin?

We live in a world of curved space, parallel universes, and collapsing wave functions; where every person spawns a million branching realities with every breath they take.

The atomic theory of matter has been completely abandoned in favor of a myriad complex of waves and fields.

It seems any problem can be solved with more ~~epicycles~~ fields!

So much for the working hypothesis that nature is simple and elegant in design.

Why is curved spacetime supposed to be more appealing than 'action at a distance'? Are they not, essentially, equally 'occult'?

No matter. It is a giant leap from saying 'in our model we imagine spacetime to be curved' to declaring emphatically that 'spacetime is curved', or that time is not real, nor motion, etc.

These ideas fly in the face of common sense, and yet people hold them fiercely.

Even the physicist who believes that time and motion are nothing but an illusion will jump out of the way of a speeding bus; one would hope!

We propose any physical theory must pass "The Speeding Bus Test".

Nature is mechanical and *dynamical*. Particles and their interactions can be completely described and explained in terms of fundamental, 'solid', and *real* units of matter, constantly in motion, and continually exchanging energy and momentum.

That is the premise of the universal model.

The Lorentz force:

Let's begin with our usual visit to the Lorentz force; a veritable new gold mine of information!

In this section, we will review what we have learned to date, and then see what other information we might extract from this equation, and what further generalizations we may make about the nature of forces and particle interactions.

The complete, 'classical', relativistic, Lorentz force between two identical electrons is

$$\mathbf{F} = (G/c^2 - (e/m_e)^2(\mu/4\pi))(c/R)^2(m_1 m_2 \mathbf{r} + (1/c^2)(\mathbf{p_1} \times \mathbf{p_2} \times \mathbf{r})) \quad (1)$$

where, of course, $\mathbf{F_1} = -\mathbf{F_2}$. Newton's second law tells us

$$\mathbf{F_1} = m_1 \mathbf{a_1} \quad (2)$$

If we compare equations (1) and (2) we can see that the acceleration of a particle is *independent of its mass*.

$$\mathbf{a_1} = \text{Function}(m_2, \mathbf{R}) \quad (3)$$

This is a general result that we used to assume applied only to the gravitational interaction.

In addition, the relative strengths of the four terms in equation (1), or the 'four forces of classical physics', are approximately as follows (superseding previous estimates ...);

electricity = 1 ; magnetism ~ $1/c^2$; gravity ~ G/c^2 ; magnetic gravity ~ G/c^4

I think Maxwell would approve!; except that we have no more need for his fields.

The two electrons exert equal and opposite forces on one another during the interaction, and the evolution of the force is completely described by the (variable) mass-energy of the two electrons, $m_1(\mathbf{r_1}(t))$, $m_2(\mathbf{r_2}(t))$, where the time, t, is *common* to both electrons.

Of course, without the fields there is no mathematical or physical mechanism to explain the interaction of these two particles 'at a distance'.

We like the idea of 'one virtual photon' (which, of course, is a discovery of the field theory model!) constantly coupling the particles; a time varying conduit for energy and momentum exchange. Can we infer or derive the virtual photon without resort to field theory?

We will defer this investigation for now. A challenge to the reader, perhaps.

Let's look at our 'new' Lorentz force in a little more detail. If we define

$$K == (G/c^2 - (e/m_e)^2(\mu/4\pi)) \quad (4)$$

then we can write equation (1) as

$$\mathbf{F} = K*(c/R)^2(m_1 m_2 \mathbf{r} + (1/c^2)(\mathbf{p_1 x p_2 x r})) \quad (5)$$

Next, we factor out the particle masses from the momentum term

$$\mathbf{F} = K*(c/R)^2(m_1 m_2 \mathbf{r} + (1/c^2)(m_1 m_2 \mathbf{v_1 x v_2 x r})) \quad (6)$$

$$\mathbf{F} = K*(c/R)^2(m_1 m_2)(\mathbf{r} + (1/c^2)(\mathbf{v_1 x v_2 x r})) \quad (7)$$

where **r** is the unit vector **R**/R.

Since our two masses form a closed, conservative system, we can 'normalize' our force by dividing by the total energy of the system; $E_{TOT} = m_1 + m_2$.

$$\mathbf{F}/E_{TOT} = K*(c/R)^2 \mu (\mathbf{r} + (1/c^2)(\mathbf{v_1 x v_2 x r})) \quad (8)$$

where $\mu(\mathbf{R}, d\mathbf{R}/dt) = m_1 m_2/(m_1 + m_2)$ is the reduced mass of the two body system.

In order to investigate the vector cross product term, we will assume our two masses (no longer necessarily electrons) are equal and orbiting one another.

Then we can write

$$\mathbf{F}/E_{TOT} = K*(c/R)^2 \mu (1 - (1/c^2)(v^2)) \quad (9)$$

$$\mathbf{F}/E_{TOT} = K*(c/R)^2 \mu - K*(\mu v^2/R^2) \quad (10)$$

We will call the second term in equation (10), the *coriolis* force, because ...

Why not? :)

If we recast our force equation into polar coordinates and allow $m_1 \neq m_2$ (i.e. for the study of planetary motion; Kepler's equations), we will pick up the usual *centrifugal* force term, in addition to our new *coriolis* force term.

Finally, there will be a force term corresponding to the interaction of the spin/angular momentum (σ, l) of one object with the 'magnetic force vector' of the other object; F_{spin}. We defer a further discussion of this idea for now.

(Remember, in our model, the relativistic mass of an object is due to the *total* relativistic motion of the mass; including spin!)

So, the total Lorentz force, for cosmology for example, will consist of four terms;

$$F_{universal} \sim F_{central} + F_{centrifugal} + F_{coriolis} + F_{spin} \qquad (11)$$

Inertial reference frames:

Because spin is an inherent component of our theory, and because everything is spinning, there can be no inertial reference frames, even in principle!

We suggest the inertial observer (who is always 'at rest') reference their inertial coordinate system to the " 'fixed background' of 'empty space' ". (You may rearrange the "scare quotes" as you'd like!)

The 'fixed stars' are no longer fixed, nor must we worry about their influencing our measurements. The stars will either be part of our study, or too far away to matter.

As to spinning buckets, we still make no hypothesis.

Life needs some mystery!

:)

The Dirac equation:

The Dirac equation for a free electron is

$$H\psi = (\alpha \cdot p + \beta m_0)\psi \qquad (12)$$

Our 'gauge invariant' solution as intimated in "A critical examination of classical and quantum mechanical waves" would look something like this;

$$\psi = \exp(-i^*m_0)\exp(-iE^*t)\exp(i\mathbf{p}\cdot\mathbf{x}) \qquad (13)$$

Replacing E(t) with the relativistic mass m(t), gives

$$\psi = \exp(-i^*m_0)\exp(-im^*t)\exp(i\mathbf{p}\cdot\mathbf{x}) \qquad (14)$$

The first factor in equation (13) is a global phase factor representing the invariant rest mass-energy/charge of the particle. This is an extra and annoying charge which is already included in the second factor of the equation; i.e the total mass-energy/charge, m(t), which we take to be our locally variable 'phase factor' *and* the locally variable interaction charge of the particle.

However, our theory is not really a gauge theory, and there seems to no formal way (e.g. by a series expansion) to have the global phase factor in equation (14) cancel the rest mass energy term in the Dirac equation.

So ... we take our lead from the standard model and throw the rest mass term away!

However, we needn't 'recover' the rest mass from an interaction with the Higgs field as in the standard model. In our model the contribution of the rest mass energy is already accounted for in the total relativistic mass energy of the particle, which is the particle's interaction charge.

To introduce interactions into the Dirac equation, we modify the standard replacement

$$p_\mu \rightarrow p_\mu + eA_\mu \qquad (15)$$

to include the gravitational interaction, and to account for the relativistic mass as the fundamental interaction charge, and obtain

$$p_\mu \rightarrow p_\mu - i^*\hbar(G^{1/2} - e/m_{rest})A_\mu\, \partial/\partial t \qquad (16)$$

This substitution should then yield the Universal Interaction Lagrangian of equation (f).

$$L\_interaction = -i \hbar (G^{1/2} - e/m_{rest})(\psi^{bar} \gamma^\mu A_\mu \, \partial\psi/\partial t) \qquad (17)$$

We interpret the component A_μ as representing a *virtual* photon, and thus we have no need for the antisymmetric tensor $F^{\mu\nu}$, and the corresponding E and B fields, to facilitate the interaction or to carry energy and momentum. Remember, in our model, all the energy and momentum in an interaction is carried by the particles!

The virtual photon is already coupled to, or 'tethered' between, two mass-charge currents, only one of which is indicated in equation (17). The second current could be a similar leptonic current or a real photon current!

In our model, we expect to treat the real and virtual photons as separate particles, each with their own 'wave function'.

Gauge theory:

Technically, our theory is *not* a gauge theory, although it does exploit and explain the observed gauge invariance of the electromagnetic interaction (another ingenuous discovery of the standard model, to be sure!).

In our model, the resolution of the 'gauge invariance issue' does not involve, or allow, the capricious variation of local charge with corresponding and compensating potential fields. (This is a feature never truly exploited, or properly explained, or *necessary* in QED ....)

Instead, in the universal model, we have a locally varying charge because the mass-energy is the coupling charge of a particle, and this varies during an interaction.

In conclusion, since the standard model explanation (and implementation) of the required 'gauge invariance' observed in QED is now thought to be incorrect, the generalization of gauge theory to explain the weak and strong interactions (thus yielding the W, the Z, and eight gluons) is probably also incorrect.

Of course, we have already removed the W, the Z, and gluons and quarks from our model.

However, the more corroborating arguments for a new theory, the better.

The quantum of action:

Our rather disdainful dismissal of the quantum of action (in "On Wave Particle Duality and the Quantum of Action", no less!), as a glorified conversion constant was both hasty and ill-advised, particularly considering our theory of the mechanical nature of the photon as an harmonically oscillating polarization, or angular momentum vector traveling at the speed of light!

What we *meant* to say was that the quantum of action, h, is the fundamental unit of matter and is inextricable, and inseparable, from particle spin.

The photon:

The photon is one "free" quantum of action. Our photon is an inertialess, massive particle of 'spin = 1' (i.e. the photon has an angular momentum, L = h). The angular momentum vector is aligned along the direction of motion of the photon. This projection of the angular momentum L along the direction of motion (the photon polarization), oscillates harmonically with a frequency, ν = E/h.

A 'plane polarized' photon switches from left handed polarization to right handed polarization every 2π radians. This 'flipping' constitutes the energy and linear momentum of the photon. The angular momentum of the photon is always equal to h.

This model is to be contrasted with the *incorrect* theory of the photon as an inertialess blob of energy that somehow "swells up" in an undefined way as it acquires more energy and momentum.

Imagine a standing, plane polarized, electromagnetic wave. The electric field oscillates harmonically exhibiting the usual characteristic nodes where the field strength goes to zero.

This standing wave is made up of photons. How do the photons manifest as an oscillating electric field?

The photon polarization behaves like a simple harmonic oscillator. Just like our particle in a box, it cannot interact as the angular momentum vector passes through the zero point. Also, when the photon is spinning in one direction, the associated electric field vector points up, when it is spinning in the other direction, the electric field vector points down.

This idea is worth repeating. A free photon cannot physically interact when the photon polarization is passing through zero. This model of the photon is able to explain the partial transmission and partial reflection of light waves incident on on a thin sheet of glass.

The neutrino:

The neutrino is one "bound" quantum of action. The neutrino has an intrinsic angular momentum of L = hbar/2. We note that

$$\text{hbar}/2 = h/4\pi \tag{18}$$

where the factor of 4π is interpreted to be the 'solid angle integral'.

Hence, we conclude the electron neutrino is one unit of intrinsic angular momentum, h, per unit volume of space.

This is all we have to say about the neutrino for now.

In our next paper, we shall try to use this observation to derive the neutrino mass, unless a 'challenged reader' beats us to the punch!

The running of alpha:

In the standard model, the electromagnetic coupling strength is expressed in terms of the electric charge, e;

$$\alpha = e^2/4\pi\varepsilon\, \text{hbar}\, c \tag{19}$$

In our model, the 'electric charge' looks like this

$$e \rightarrow (m*e/m_e) \tag{20}$$

and alpha becomes

$$\alpha = m^2(e/m_e)^2/4\pi\varepsilon\, \text{hbar}\, c \tag{21}$$

$$\alpha = (m_e^2/1 - v^2/c^2)(e/m_e)^2/4\pi\varepsilon\, \text{hbar}\, c \tag{22}$$

$$\alpha = (1/1 - v^2/c^2)\, e^2/4\pi\varepsilon\, \text{hbar}\, c \tag{23}$$

$$\alpha = \alpha_0(1 + (v/c)^2 + (v/c)^4 + \ldots) \tag{24}$$

In our running of alpha, the velocity squared replaces the four-momentum transfer, Q^2, and there is no need to introduce an arbitrary cut-off mass.

Magnetic moments:

The standard model formula for the magnetic moment of the electron is

$$\mu_e = e\hbar/2 m_e \quad (25)$$

where, it seems, m_e is the *rest mass* of the electron. This formula can be 'derived' from some fairly dubious hand waving arguments, and, of course, it "falls out naturally" from the Dirac equation.

In a previous paper, we were curious about the behavior of the magnetic moment at relativistic speeds. We naively replaced the rest mass, m_e, with the relativistic mass in equation (25). The results did not make sense (although we tried to 'make hay' with them anyway!)

The electron magnetic moment decreased at relativistic speeds. This seemed to run counter to common sense, *and* to our theory that the electron magnetic moment arises from spinning mass, and our theory that the electron "spins faster" when it is accelerated.

Since our electromagnetic coupling charge is $e \rightarrow (m \cdot e/m_e)$, we will make the same substitution as we made in our study of the running of alpha.

$$\mu_e = (m/m_e)(e\hbar/2 m_e) \quad (26)$$

Similarly, in our model, the gravitational coupling charge is considered to be $G^{1/2} m$, so our naive guess for the neutrino magnetic moment would now be

$$\mu_\nu = G^{1/2}(m)(\hbar/2 m_\nu) \quad (27)$$

Unfortunately, this formula is not dimensionally correct, so we add a factor of $4\pi\varepsilon$ 'by hand'.

$$\mu_\nu = (G/4\pi\varepsilon)^{1/2}(m)(\hbar/2 m_\nu) \quad (28)$$

It looks like this part of the model still needs some work.

A challenge to the reader! :)

Conclusion:

Our new model is now 'complete'.

I hope the physics community will adopt it, formalize everything,
and whip it into proper shape.

I reckon the results should also be compared with existing data, just as a check!

As far as new experiments are concerned, the structure function results from the Jefferson Lab collaborations seem quite intriguing. These measurements should be examined in light of our new theory of the proton as a bound state of two positrons and an electron.

Elsewhere, the world over, people are devising new neutrino experiments.

These experiments are exactly what we need just now to confront the basic premise(s) of new our model.

Serendipity!

Spooky.

Summary:

Here's a 'back of the envelope summary' of the universal model (to date!), including results from several of our previous papers.

The Lorentz force:

$$\mathbf{F} = (G/c^2 - (e/m_e)^2(\mu/4\pi))(c/R)^2(m_1 m_2 \mathbf{r} + (1/c^2)(\mathbf{p_1} \times \mathbf{p_2} \times \mathbf{r}))$$

The QED Lagrangian:

$$L_{interaction} = -i*\hbar(G^{1/2} - e/m_{rest})(\psi^{bar} \gamma^\mu A_\mu \, \partial\psi/\partial t)$$

The running of alpha:

$$\alpha = \alpha_0(1 + (v/c)^2 + (v/c)^4 + \dots)$$

alpha_G:

$$\text{alpha\_G} = m\_e^2 * G/(\hbar * c) \quad ; \; m\_e \text{ is the relativistic mass.}$$

alpha_weak:

$$\text{alpha\_weak} = m\_v^2 * G/\hbar * c \quad ; \; m\_v \text{ is the relativistic mass.}$$

alpha_strong:

Please consult "The Sinister Universe".

The nucleons:

$$p = (e^+, e^+, e^-) \; ; \; n = (e^+, e^-, \bar{\nu}_e)$$

The pion:

$$\pi^0 = 1/(2)^{1/2} (e \, \bar{e} - \mu \, \bar{\mu})$$

The Higgs boson:

$$H = (\nu_e, \bar{\nu}_e)$$

The leptonic table:

LEPTONS   ANTI-LEPTONS

electron	electron neutrino	PARITY ⇔	electron antineutrino	positron
⇐	CHARGE	MASS ↕	CHARGE	⇒
muon	muon neutrino	PARITY ⇔	muon antineutrino	anti-muon
⇐	CHARGE	MASS ↕	CHARGE	⇒
tau	tau neutrino	PARITY ⇔	tau antineutrino	anti-tau
⇔	weak isospin	mass isospin ↕	weak isospin	⇔

TABLE 1: The leptons and their interrelations; or the kleptogenesis of the leptoquarks.

Any lepton can be 'generated' from any other by the appropriate applications of the parity operator, the weak isospin operator, and our newly proposed 'mass isospin' operator.

Muon decay:

The muon "sheds" a muon neutrino to become a "generic" virtual charged lepton.

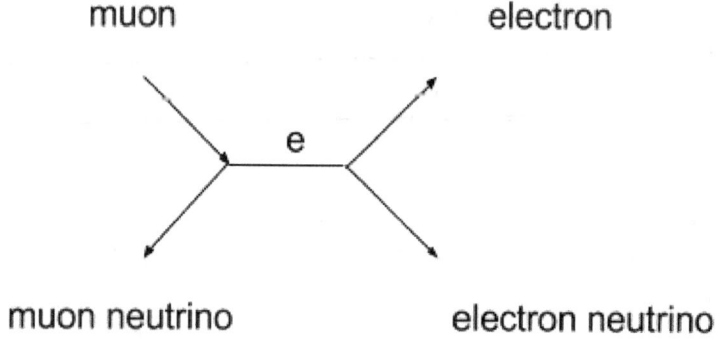

FIGURE 1: Muon decay. The propagator is most likely a generic virtual lepton, e/mu/tau.

References:

Modern Elementary Particle Physics
Gordon Kane

Classical Dynamics of Particles and Systems
Jerry B. Marion

Foundations of Electromagnetic Theory
John R. Reitz, Frederick J. Milford, Robert W. Christy

Quantum Physics
Rolf G. Winter

Gauge Theories in Particle Physics
I. J. R. Aitchison and A. J. G. Hey

Quarks and Leptons: An Introductory Course in Modern Particle Physics
Francis Halzen, Alan D. Martin

Quantum Field Theory
F. Mandl, G. Shaw

Theoretical Mechanics of Particles and Continua
Alexander L. Fetter, John Dirk Walecka

and

Elementary Modern Physics   (Best Book Ever!)
Richard T. Weidner, Robert L. Sells

Books by Greg Feild:

1. "A quantum mechanical theory of gravitational interactions"
   CreateSpace Independent Publishing, 8/29/2016

2. "Observations on the quantum mechanical nature of gravity"
   CreateSpace Independent Publishing, 10/8/2016

3. "On gravitation and electric charge"
   CreateSpace Independent Publishing, 10/29/2016

4. "On spin, mass, and charge"
   CreateSpace Independent Publishing, 11/29/2016

5. "On angular momentum, acceleration, and absolute motion"
   CreateSpace Independent Publishing, 1/1/2017

6. "The Sinister Universe"
   CreateSpace Independent Publishing, 3/1/2017

7. "On Parity and Isospin"
   CreateSpace Independent Publishing, 4/11/2017

8. "Reflections on the Sinister Universe"
   CreateSpace Independent Publishing, 5/12/2017

9. "On Current Physics"
   CreateSpace Independent Publishing, 6/11/2017

10. "A Critical Examination of Classical and Quantum Mechanical Waves"
    CreateSpace Independent Publishing, 6/18/2017

11. "On wave particle duality and the quantum of action"
    CreateSpace Independent Publishing, 7/6/2017

Compilation:

"The Universal Model of Our Sinister Universe: The First Ten Books"
CreateSpace Independent Publishing, 7/2/2017

Notes: :)

Secret song:

Or … we could call it the **SUM** (total of everything!) model.

The **S**inister **U**niverse **M**odel

Or … we could just call it greg's model.

Grass **R**ooted, **E**mergent **G**ravity and **S**pace

grass root physics!

On Action and Reaction

Greg Feild

September 24, 2017

The final test of deduction lies in experimental observation.

Elaboration by reasoning may make a suggested idea very rich and very plausible, but it will not settle the validity of that idea.

Only if facts can be observed (by methods of collection or experimentation), that agree in detail and without exception with the deduced results, are we justified in accepting the deduction as giving a valid conclusion.

Thinking, in short, must end as well as begin in the domain of concrete observations, if it is to be complete thinking.

-- John Dewey
How We Think

Abstract:

In this penultimate paper on the universal model of our world,
we take stock, and examine the current status of our model in
the spirit of our two previous efforts; "On Spin, Mass, and Charge"
and "The Sinister Universe".

We will refrain from (*too*) much new speculation,
so as to not muddy the waters.

However, we will investigate several of the unresolved issues and new
proposals highlighted in our last paper, "On Matter, Mass, and Motion".

About the author:

Greg Feild is a 'gentleman' scientist
and an armchair philosopher.

His goal is to make physics fun again ...
and physical!

Coming soon:

"A Quantum Mechanical Theory of Everything" !

:)

Preface:

Our new universal model of the world actually consists of many separate, and seemingly disparate new theories, that are however, logically intertwined; both suggesting and reinforcing one another. It basically explains every outstanding problem of the standard model and more! A bold statement to be sure.

Having started our enquiries 'from scratch', with a broad view of the successes and perplexities of all current models, we were able to confront and dispose of these problems, taking one at a time, with 'relative' ease.

Bold as love!   :)

Now, whether our new model is asymptotically approaching coherence and/or completeness, is another question, and one, ultimately, not to be decided by ourselves!

In the brief synopsis that follows, we assume the reader is a fan (or frenemy) of our new model, or is at least familiar with our previous 12 books.

The 'universal model' can be broadly characterized, or divided into, five complementary (and traditionally separate or distinct) 'sub-theories';

1) A quantum and cosmological theory of gravitational interactions
2) A classical and quantum mechanical theory of all particle interactions
3) A theory of time and space
4) A theory of the fundamental nature of elementary particles
5) A physical interpretation of the quantum mechanical wave function

In our model, gravitation appears to be a 'carbon copy' of electromagnetism, except for the lack of a negative gravitational charge.

Otherwise, gravity and electromagnetism differ only by a factor of *scale(s)*
(i.e. by factors of G, $e/4\pi\varepsilon$, and $1/c^2$),

The fundamental coupling charge, for both quantum and cosmological interactions, in both gravitational interactions *and* electromagnetic interactions, is the *total* relativistic mass of the particle or planet involved.

From this realization alone, most of the theory falls almost magically into place.

We now present a 'laundry list' of the issues we believe are explained by the model:

- Matter-antimatter asymmetry
- The nature of antimatter
- Matter-antimatter annihilation
- The origin of parity
- Neutrino handedness
- The nature of the strong force
- Why the proton charge is exactly e
- Parton confinement and asymptotic freedom
- The weak force
- The origin of mass and charge
- Muon decay
- Particle families
- Gravity
- The $1/R^2$ law
- Wave particle duality
- The wave function
- The nature of the photon
- The nature of leptons
- The spinor nature of the electron
- SU(3)
- Gauge theory
- The running of alpha
- Beta Decay
- Why the Born approximation is exact
- Issues in cosmology
- Relativistic mass
- Dark matter
- Dark energy
- Quantization
- And much, much, more !

Introduction:

In our first paperbook "A Quantum Mechanical Theory of Gravitational Interactions", we began our enquiries by proposing a modified principle of equivalence;

inertial mass = gravitational mass = gravitational charge   (1)

Our recent investigations suggest an even *more* general expression for this principle, which we shall call the 'universal principle of equivalence'.

relativistic inertial mass = relativistic gravitational mass = gravitational charge

== *the fundamental universal coupling charge*   (2)

In our new model, the "relativistic" particle mass is the gravitational charge, the "strong force" charge, the weak charge, and ultimately, the electric charge as well!

The universal theory of relativity:

Why do we use "scare quotes" when we say the 'relativistic' mass?

The postulates of the special theory of relativity are;

i) Physical laws are the same in all inertial reference frames.
ii) The speed of light is a universal constant in any reference frame.

As an alternative, we propose the postulates of 'the *universal* theory of relativity';

i) Particle interactions occur at the speed of light in any reference frame.
ii) Particle interactions obey the $1/R^2$ law in any reference frame.

iii) $F = dp/dt$

In an inertial reference frame, I believe, these three postulates should lead one (the challenged reader!) to the familiar expressions for the relativistic mass and momentum, *without* reference to, or resorting to, coordinate transformations.

The familiar Lorentz transformations would then follow naturally as a consequence.

The speed of light:

All particle interactions occur at the speed of light. Photons are interacting particles that travel at the speed of light. Photons have the same speed in all inertial reference frames.

However, photons do not have the same energy, or frequency, in all inertial reference frames.

This last point is 'obvious', but it also seems very important somehow …

We will use this observation to establish a universal reference frame, absolutely at rest; a fixed background that is always 'at rest' relative to any and all motion.

The universal reference frame:

An inertial observer is *defined* to be absolutely at rest, against the fixed background of space, if they measure the accepted, absolute value for a particular and well known wavelength of the cosmic background radiation.

An observer in a *moving* inertial reference frame, could then determine their absolute velocity relative to the fixed background of space, by measuring the Doppler shift of the cosmic microwave background.

The cosmic microwave background:

So, what is the cosmic microwave background?

We've essentially ruled out "stretched" photons due to either, a) the big bang, or b) an eternally expanding and contracting universe.

Our current model suggests a static and eternal universe.

So, the cosmic microwave background must be gravitational bremsstrahlung!

A collection of soft photons, emitted by every massive particle, ever accelerated, ever!

The Lorentz force:

In "On Matter, Mass, and Motion', we found it convenient to express the Lorentz force between two bodies as a function of the total energy of the system;

$$F/E_{TOT} = K*(c/R)^2 \mu (\mathbf{r} + (1/c^2)(\mathbf{v_1 x v_2 x r})) \tag{3}$$

where $\mu(\mathbf{R}, d\mathbf{R}/dt) = m_1 m_2/(m_1 + m_2)$ is the reduced mass of the two body system, and $\mathbf{r} = \mathbf{R}/R$. The reduced mass, is the relativistic mass, and depends on the position *and* velocity of the particles. The constant K consolidates all the coupling constants;

$$K == (G/c^2 - (e/m_e)^2(\mu/4\pi)) \tag{4}$$

We interpret the second term in equation (3) as a generalization of the familiar coriolis force which arises when studying the motion of bodies on earth from a fixed reference frame.

In our generalized Lorentz force, the coriolis force is not an artifact of the choice of a particular reference frame, but arises from the absolute relative motion of the two bodies.

In our model, the coriolis force is *real*, because all forces are velocity dependent due to the fact that particle mass is velocity dependent; $m = m(\mathbf{r},\mathbf{v})$.

Similarly, the centrifugal force is now also a real force and is not due to "space time" disturbances as in the general theory of relativity

Equation (3) should be 'easy' to generalize to an N body system.

The Lorentz torque:

The 'Lorentz torque' arises from the interaction of the intrinsic angular momentum (quantum spin or classical moment of inertia) of one body with the magnetic field vector of the second body, yielding the force F_{SPIN} introduced in "On Matter, Mass, and Motion".

We still aren't ready to work out the formula for F_{SPIN} (readers?), but we do note that the magnetic force *now does work*, and this force is equal, opposite, and *central* between the two bodies.

It seems the extra, work free, looping and spiraling, circular motion of a charged particle in an external magnetic field is one of Nature's red herrings! Artful window dressing.

The classical propagator:

In our last book, we challenged the reader (and ourselves!) to create a 'propagator' from our new universal Lorentz force.

The complete, 'classical', relativistic, Lorentz force between two identical electrons is

$$F = (G/c^2 - (e/m_e)^2(\mu/4\pi))(c/R)^2(m_1 m_2 \mathbf{r} + (1/c^2)(\mathbf{p}_1 \mathbf{x} \mathbf{p}_2 \mathbf{x} \mathbf{r}))$$

where, of course, $F_1 = -F_2$. We shall start our study slowly, and consider only the static, or Coulomb potential, as one usually does.

$$F_{1,2} = K*(c^2/(r_1 - r_2)^2(m_1(r_1, v_1))(m_2(r_2, v_2))$$

The intriguing factor of c^2/R^2 (the result of some factoring), already looks 'propagator like', with units of $1/t^2$.

We could look at the work

$$W_{1,2} = -W_{2,1} \Rightarrow W_{1,2} + W_{2,1} = 0$$

or even the action. I believe the key to this problem is converting the integrals such that we are integrating over the masses, dm, of the two particles; the limits of integration being the initial and final relativistic masses of the two particles.

We even think we have a proof, but unfortunately it will not fit in this giant space below.

The electron:

In "On Matter, Mass, and Motion", we derived/induced the universal model generalization of the formula for the magnetic moment of the electron;

$$\mu_e = (e/m_e)(m)(\hbar/2m_e) \qquad (5)$$

That is

$$\mu_e = (\text{coupling constant})*(\text{mass})*(\text{angular momentum per unit mass})$$

So, the electron is one unit of angular momentum per unit mass per unit volume of space; or one unit of inertial, half integral, angular momentum per unit mass.

$$e \Rightarrow h/4\pi m_e == (\hbar/2)/m_e == L/m_e \qquad (6)$$

Inserting the formula for the relativistic mass into equation (5) we get

$$\mu_e = (e*\hbar/2m_e)(1/(1 - v^2/c^2)^{1/2}) \qquad (7)$$

We then make the usual series expansion to obtain

$$\mu_e = (e*\hbar/2m_e)(1 + \tfrac{1}{2} v^2/c^2 + \tfrac{3}{8} v^4 c^4 + \ldots) \qquad (8)$$

and find the magnetic moment of the electron increases with velocity, *as expected*.

The electron is essentially the vector L/m_e. This vector precesses about the axis of the direction of motion with a frequency; $\nu = E/h = m/h$. As the speed of the electron increases, the frequency of precession increases, and the mass of the electron increases.

The surprising 'spinor' nature of the electron is due to the angular momentum vector flipping helicity/polarization every 2π radians.

Even though the mass and magnetic moment of the electron increase with the velocity, the electron angular momentum is always $\hbar/2$.

We finally know what has been waving all this time!

The waving of the wave equation/wave function represents the periodic precession of the electron spin about the direction of travel of the electron!

N.B. The wave equation describes *many* physical phenomena that don't really wave.

Spinoring:

We've been saying the "helicity or polarization" of the electron 'flips' every 2π radians.

Technically, of course, this terminology is incorrect, because even though there *is* 'flipping' going on, the helicity of the electron never changes!

Instead, we shall choose to say the electron is 'spinor-ing', as illustrated in Figure 1.

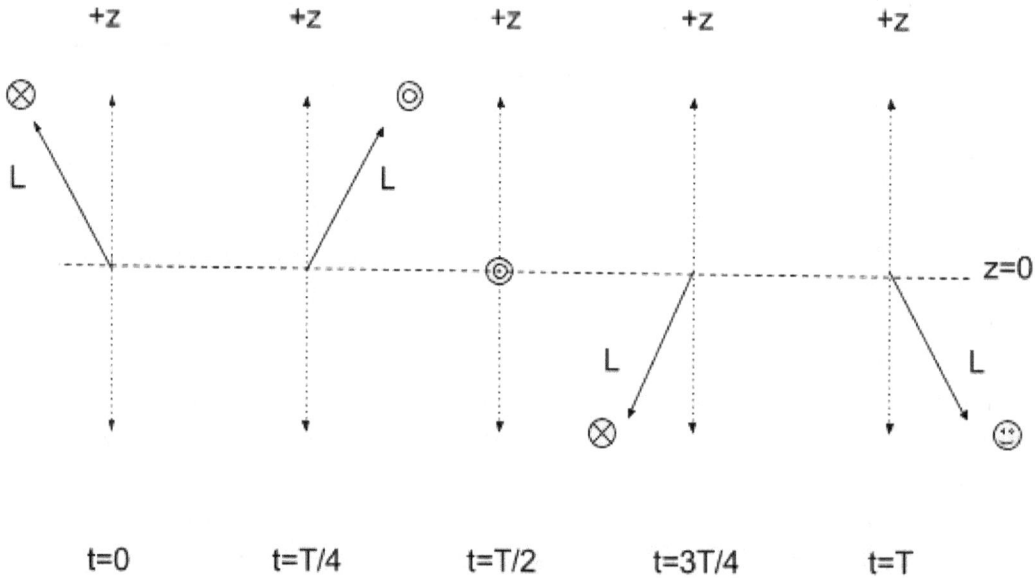

Figure 1: At rest with an electron traveling the the z-direction. The spin angular momentum vector 'precesses' about the direction of motion, tracing out a closed, three dimensional figure eight (a string!). The x symbol represents motion into the page. The dot symbol represents motion out of the page. At time T/2, we see the the angular momentum is *perpendicular* to the direction of travel. (This is when the electron engages in 'virtual' interactions.)

We also can see, that although the angular momentum vector is 'spinor-ing', the polarization, or helicity, of the electron does *not* change, and is constant!

Whew!

The neutrino:

We believe (really, really want) the electron neutrino to have a magnetic moment. Actually, we are (almost) *sure* it does, and we are determined to figure out!

This hypothesis is a core premise of our model! So, consider the electron ...

Even though the electron is electrically charged, we can see from equations (5) and (8), that the magnetic moment is really only dependent on the mass of the electron.

The electric charge, e, functions solely as a proportionally factor or *coupling constant*.

Of course, this has been our premise since "On Gravitation and Electric Charge".

Now, we will construct the neutrino magnetic moment in strict analogy with equation (5)

$$\mu_\nu = (m)(\hbar/2m_\nu) \qquad (9)$$

Originally, way back in "Observations on the Quantum Mechanical Nature of Gravity", we expected or assumed the neutrino magnetic moment would have the same units, or "charge", as the electron magnetic moment, involving the electric charge, e.

In hindsight, we can see this approach was wrong for two reasons

1) The gravitational magnetic vector term, **B_g**, of our new Lorentz force does not depend on the electric charge.
2) We have already checked that the 'cross term' in our expression for alpha_strong, as derived in "The Sinister Universe", is dimensionally correct.

If we take equation (9) as the magnetic moment of the neutrino and form the vector dot product with either **B** or **B_g** from our generalized Lorentz force, we see the units also work out correctly.

So, we have *finally* figured out the magnetic moment of the neutrino!

Whew.

(We may have just achieved complete coherence!)

Inserting the formula for the relativistic mass of the neutrino into equation (9) we see

$$mu\_v = (hbar/2)(1 + \tfrac{1}{2} v^2/c^2 + \tfrac{3}{8} v^4 c^4 + ...) \qquad (10)$$

And, we must conclude that *the neutrino is one inertial quantum of action* as we proposed in "On Matter, Mass, and Motion". So, we can represent the neutrino symbolically, as we did for the electron in equation (6), and write

$$nu\_e \Rightarrow hbar/2 == L \qquad (11)$$

Actually, equation (10) shows that *there is only one neutrino*. The three neutrinos differ only by their **velocity.**

The tau lepton is the most massive of the three charged leptons, so when it 'decays', it emits the highest velocity, and hence, most massive of the the "three" neutrinos, usw.

The tau is shedding energy and spin, because what else is there?

We can also derive the magnetic moment of the neutrino using the hand waving arguments usually assumed for the electron magnetic moment.

$$mu = (mass\ current)(area) = (m/t)(A) = m(L/2m) = hbar/2 \qquad (12)$$

Finally, we reach way back to Physics 101 and recall the formula for angular momentum

$$L = mvr \qquad (13)$$

The angular momentum of the neutrino is h and we assume it spins with angular velocity c.

$$h = (m\_nu)(r\_nu)c \qquad (14)$$

Now, we can solve for the *Compton radius of the neutrino*; the smallest probable distance.

$$r\_nu = h/(m\_nu)*c \qquad (15)$$

The photon:

The photon is essentially a perpetual motion machine! The photon polarization oscillates harmonically at a frequency proportional to its energy, nu = E/h, as discussed in "On Matter, Mass and Motion". This is an unorthodox picture of the photon polarization, but we note it satisfies the orthogonality condition on the photon wave vectors; $\mathbf{k} \cdot \varepsilon = 0$.

The photon is one 'free', massive, but inertialess, unit of angular momentum; L = h.

The neutrino is one 'bound', massive, unit of angular momentum *per unit space*; L = hbar/2.

The electron is one 'bound' unit of angular momentum per unit space *per unit mass*; L = hbar/2.

Particle interactions:

Take a look at the angular momentum vector of the electron at time, t=0, in Figure 1. Here, we say the electron helicity is in 'full bloom' and it is able to absorb a real photon (*if* its polarization is also 'blooming'), increasing the rate of precession of the electron angular momentum vector, and hence the electron mass.

Consider the case of partial transmission and partial reflection of light from a thin sheet of glass. Photons in 'full bloom' will be transmitted. Photons with polarization ~0 will be reflected.

Similar arguments can be made for electron tunneling. If an electron arrives at a potential barrier 'out of phase for reflection', *and,* the electron 'wavelength' is comparable to the 'height' of the potential barrier, the electron will 'tunnel' through!

Quantum field theory:

In our model, particles are not created and destroyed. Instead, particles absorb, emit, merge with, and *shed* one another.

For example, an electron and a positron do not 'annihilate', producing a virtual photon.

The electron and positron have equal and opposite spins, ½. They merge to become a photon of spin 0. The photon then splits into the particle antiparticle pair demanded by the situation.

In our model, particle interactions are a continuous flux and flow of energy and momentum, *flowing only one way*; futureward. Particles interact by exchanging units of *angular momentum*.

13

Conclusion:

I apologize for all the homework!

However, building a model is a collaborative effort, as we've emphasized before.

:)

And, math is hard! It would take me a very long, frustrating while to (try and) work through all the math we have suggested. So, I leave it to the professionals!

I would much rather get the core ideas out to the community, than try to be a 'hero' and work through all the math myself.

There may be still some 'issues' with the model, of course, but only one person has been working on it, and for only one year! We need more people power and hours.

I believe this model, and it's several theories, is essentially correct.

I *know* there are plenty of good bits.   (Maybe even some 'clickbait' !)

Unfortunately, I am an 'outsider' (even though I have a PhD in physics, twenty years experience in the field, am the author of many conference proceedings, and the principal author of at least 5 physics papers published in peer reviewed journals, and the author of literally thousands of papers as the member of two international experimental collaborations).

But, rant over.   :)

Someone will discover my little books.

Someday.

They will be *my* hero!

Outsider physics!

home grown

References:

Modern Elementary Particle Physics
Gordon Kane

Classical Dynamics of Particles and Systems
Jerry B. Marion

Foundations of Electromagnetic Theory
John R. Reitz, Frederick J. Milford, Robert W. Christy

Quantum Physics
Rolf G. Winter

Gauge Theories in Particle Physics
I. J. R. Aitchison and A. J. G. Hey

Quarks and Leptons: An Introductory Course in Modern Particle Physics
Francis Halzen, Alan D. Martin

Quantum Field Theory
F. Mandl, G. Shaw

Theoretical Mechanics of Particles and Continua
Alexander L. Fetter, John Dirk Walecka

and

Elementary Modern Physics   (Best Book Ever!)
Richard T. Weidner, Robert L. Sells

a feild theory		:)

Books by Greg Feild:

1. "A quantum mechanical theory of gravitational interactions"
   CreateSpace Independent Publishing, 8/29/2016

2. "Observations on the quantum mechanical nature of gravity"
   CreateSpace Independent Publishing, 10/8/2016

3. "On gravitation and electric charge"
   CreateSpace Independent Publishing, 10/29/2016

4. "On spin, mass, and charge"
   CreateSpace Independent Publishing, 11/29/2016

5. "On angular momentum, acceleration, and absolute motion"
   CreateSpace Independent Publishing, 1/1/2017

6. "The Sinister Universe"
   CreateSpace Independent Publishing, 3/1/2017

7. "On Parity and Isospin"
   CreateSpace Independent Publishing, 4/11/2017

8. "Reflections on the Sinister Universe"
   CreateSpace Independent Publishing, 5/12/2017

9. "On Current Physics"
   CreateSpace Independent Publishing, 6/11/2017

10. "A Critical Examination of Classical and Quantum Mechanical Waves"
    CreateSpace Independent Publishing, 6/18/2017

11. "On wave particle duality and the quantum of action"
    CreateSpace Independent Publishing, 7/6/2017

12. "On Matter, Mass, and Motion"
    CreateSpace Independent Publishing, 9/14/2017

Compilation:

"The Universal Model of Our Sinister Universe: The First Ten Books"
CreateSpace Independent Publishing, 7/2/2017

a greg feild production

↺ ↻
gf

Notes:

There was an old lady who ate a fly:

There was an old lady who ate a horse.

She died, of course!

an allegorical tale

a parable for our times!

keep it simple!

physics is fun!!

:)

A Quantum Mechanical Theory of Everything

Greg Feild

November 5, 2017

About the author:

Greg Feild is a physicist.
His mission is to put the *physical* back into physics.

His motivation is to understand the real world.

I suppose it is tempting, if the only tool you have is a hammer, to treat everything as if it were a nail.

-- Abraham Maslow
The Psychology of Science

Abstract:

In this book, we offer a final account of
The Universal Model of Our Sinister Universe.

The main purpose is to summarize the theory without the distraction of the numerous errors committed along the way.

This book is not an exegesis or exposition, but a synthesis of our several speculations spanning thirteen recent papers, for a final, wart free, theory of everything!

Submitted for your approval ...

Sapere aude!   -- Horace

Time is the formal a priori condition of all phenomena whatsoever. Space, as the pure form of external intuition, is limited as a condition a priori to external phenomena alone.

If I can say a priori, "all outward phenomena are in space, and determined a priori according to the relations of space", I can also, from the principle of the internal sense, affirm universally, "all phenomena in general, that is, all objects of the senses, are in time, and stand necessarily in relations of time."

-- Immanuel Kant
Critique of Pure Reason

Preface:

This book contains (only a) few new ideas and not a lot of new text.

Most of the text has be been recycled from previous books.

The purpose of this book is to provide a summary of our new model, separating the wheat from the chaff.

Please enjoy!

                                                                Greg F.

Feeling Gravity's Pull:

Reason had harnessed the tame
Holding the sky in their arms

Gravity pulls me down

                                          -- R.E.M

The universal model:

Until our new collection of theories, people did not understand the origin of particle mass, why mass and energy are equivalent, how accelerated particles acquire mass-energy, the nature of antimatter, the nature of particle-antiparticle annihilation, particle spin, the particle family hierarchy, muon decay, gravity, nor last and what *should be the very least*; the fundamental physical mechanisms underlying particle interaction.

Yet theorists speculate wildly. Experimentalists look for "new physics" !

Today's theoretical approach is all about mucking around with mathematical models; without an underlying physical model, it seems. If you get the math right, the correct model is sure to follow! But whence this 'math'? Monkeys at typewriters spring to mind ...

Physics is mathematical because we measure things and then try to organize and understand them in a quantitative way.

First you do the physics, then you do the math. Lather. Rinse. <u>Repeat</u>.

If you can *later* formalize your math into something beautiful. Great!

Lately, physicists have been putting the cart before the horse.

Actually, the cart has been abandoned in a ditch!

I *only* scold because certain physicists have been *shamelessly* airing our dirty laundry, *and ignorance*, turning a tidy profit, all the while; probably in the name of 'public education and outreach'. Unfortunately, they speak *complete nonsense*, undermining public confidence in science in general, and providing needless fodder for the fanatic followers of the latest fads and fashions.

Who is in charge of these people ?

I'm not angry anymore!   :)

Just disappointed.

I'm off to see my psychic astrologer to have my quantum hologram unentangled.

Introduction:

Scientific theories must be rational and they must be logical. Scientific theories must conform to human reason and to common sense.

Reason is all we have to discern truth from falsehood; reality from fantasy; fact from fiction.

Curved spacetime, extra dimensions, extra universes, parallel worlds, collapsing wave functions, non-locality; none of these ideas can be reconciled with human reason or sense.

For this reason alone, they must be dismissed out of hand and with extreme prejudice.

These ideas are silly.

It is the 21st century. There is no room for magical thinking, supernatural entities, or superstition in science.

Space is three dimensional. Elementary particles are solid units of matter. All interaction is mechanical and deterministic. There is only one universe.   Time flows forward.

When did humankind lose its way?

When did the physicist become ontologist?
Who allowed such a transgression?

Where were the *philosophers*?

Will human beings ever mature?
Will human nature ever change?

Or, will there *always* be beasties?

Oy vey.

The special theory of relativity:

The erroneous argument and conclusion of the special theory of relativity may be stated as:

i) Since transformations between inertial reference frames are no longer Galilean,

ii) Space and time must no longer be Newtonian.

Expressed in this manner, the argument looks weak, if not fallacious, already.

The inertial reference frames between which we must make non-Galilean transformations are themselves Newtonian reference frames, each and every one. Time and space appear the same in all inertial reference frames; flat, isotropic, homogeneous, etc. Despite its name, the special theory of relativity implicitly prefers the observer at rest and makes a special case of the observer in motion. Of course, the whole point is the observer "does not know" he is motion (without reference to something else), and time and space certainly "don't know" whether an observer is in motion. Only *relative motion between interacting particles* matter.

Choice of reference frame is a matter of convenience in bookkeeping. Space is space.

(Why figure in base twelve when you have ten fingers and toes?)

The correct argument and conclusion from the special theory of relativity is as follows:

i) Since the mass and energy of a particle are equivalent, and the energy of a particle is a *nonlinear* function of particle velocity,

ii) We may no longer make Galilean transformations between inertial reference frames.

Time and space "don't know from" your coordinate transformations!

Oy vey ist mir.

The universal principle of equivalence:

In our first book "A Quantum Mechanical Theory of Gravitational Interactions", we began our enquiries by proposing a modified principle of equivalence;

inertial mass = gravitational mass = gravitational charge (1)

Our recent investigations suggest an even more general expression for this principle, which we shall call the 'universal principle of equivalence'.

relativistic inertial mass = relativistic gravitational mass = gravitational charge

== *the fundamental universal coupling charge* (2)

In our new model, the "relativistic" particle mass is the gravitational charge, the "strong force" charge, the weak charge, and ultimately, the electric charge as well!

The universal theory of relativity:

Why do we use "scare quotes" when we say the "relativistic" mass?

The postulates of the special theory of relativity are;

i) Physical laws are the same in all inertial reference frames.
ii) The speed of light is a universal constant in any reference frame.

As an alternative, we propose the postulates of 'the *universal* theory of relativity';

i) Particle interactions occur at the speed of light in any reference frame.
ii) Particle interactions obey the $1/R^2$ law in any reference frame.

iii) $F = dp/dt$

In an inertial reference frame, I believe, these three postulates should lead one to the familiar expressions for the relativistic mass and momentum, *without* reference to, or resorting to, coordinate transformations.

The familiar Lorentz transformations would then follow naturally as a consequence.

Inertial reference frames:

In our model, the total relativistic mass of an interacting object is a function of *all* relative velocities between the object and a second interacting object.

The total 'relativistic' velocity between two bodies now includes contributions from angular velocity, and 'intrinsic', or absolute, *rotational velocity*, in addition to the usual rectilinear velocity.

Because spin is an inherent component of our theory, and because everything is spinning, there can be no inertial reference frames, even in principle!

We suggest the inertial observer (who is always 'at rest') reference their inertial coordinate system to the " 'fixed background' of 'empty space' ". (You may rearrange the "scare quotes" as you'd like!)

The 'fixed stars' are no longer fixed, nor must we worry about their influencing our measurements. The stars will either be part of our study, or too far away to matter.

The universal reference frame:

In our model, (the background of) space is fixed and immutable; 'flat', homogeneous, isotropic; Euclidean. Empty. Void.

An inertial observer is *defined* to be absolutely at rest, against the fixed background of space, if they measure the accepted, absolute value for a particular and well known wavelength of the cosmic background radiation.

An observer in a *moving* inertial reference frame, could then determine their absolute velocity relative to the fixed background of space, by measuring the Doppler shift of the cosmic microwave background.

In our model, the cosmic microwave background consists of gravitational waves (photons). Residue from every mass ever accelerated, anywhere (within "range" of our 'sector'), ever!

The speed of light:

The speed of light is the speed of particle interactions. The speed of interaction between particles is independent of the velocity of the particles, and if we feel compelled to introduce an observer, the choice of reference frame. This attests to the fixity of space.

However, the *frequency* (and thus 'strength') of particle interactions is dependent on the *relative velocity* of the particles. This *is* physics.

Time may be not be absolute, but in any given inertial reference frame, time is fixed and invariable, ticking away regularly, ceaselessly, and eternally. (i.e. Newtonian.)

This is all that matters, and all we can ask for!

Why can't a particle travel faster than the speed of light?

A particle cannot exceed the speed of light because it cannot exceed the speed of the force (or the source) causing it to accelerate!

Classically, we can imagine placing an electron in a constant electric field of infinite extent. This field would exert a constant force on the electron and, in principle, we could accelerate the electron to any speed we'd like.

However, fields are not real.

Particle acceleration is due to particle interactions. Our electron's acceleration is actually caused by the exchange of virtual photons with electrons on the surface of a capacitor plate.

The faster and farther our electron is accelerated, the farther the next virtual photons have 'to travel' to give our electron its next boost. The $1/R^2$ law in action!

We can employ similar arguments to explain why all observers measure the same value for the speed of light. All particle interactions occur at the speed of light in any reference frame.

Only the fact that the different observers measure different frequencies for said light, is keeping our minds from totally exploding right now!

Fields:

In our new universal model of the world, there is one fundamental force responsible for all particle interactions.

We continue to call this one, single, elementary, force electromagnetism, and the corresponding field the electromagnetic field.

Our new electromagnetic field interacts with the 'total coupling charge' of a particle; the sum of a 'mass dependent' electric charge and the relativistic mass of the particle, with the appropriate proportionality factors (e.g. alpha, G, etc.) applied to each term.

In our theory, the electromagnetic field is solely a mathematical field describing the interaction between particles, and *does not* carry energy or momentum.

That is the job of the photon; be it virtual or real.

It does not make sense (or, at least, it is not operationally useful) to talk about the force field or potential field of a single particle; its strength, how far it extends in space, etc.; without the presence of at least one other particle somewhere in the universe!

In the field model, a two particle universe would have an infinite amount of energy all stored in these magical fields, and all arising from two little electrons.

We conclude that a particle need not, and does not, create a force field somewhere where some other particle is not, and in fact may never be.

People get so caught up in their mathematical models that they totally *forget* the basic underlying, original, and guiding thoughts and principles that inspired them.

The first rule, and overarching principle of the universal model is - you cannot do physics with just one particle!

For every action there is an equal and opposite reaction.

Classical field theory essentially throws out Newton's third law, especially when one object is considered a fixed center of force for one or several other objects. The fixed center does not move, or 'recoil'. In addition, all other objects are considered as 'passive' participants in the interaction. Finally, the time delay between any change in the source (e.g. if it were allowed to move) and the subsequent motion of the objects subject to the force, renders the notion of action and reaction logically impossible. Most people are (at least implicitly) aware of these assumptions, but they do not realize just how much damage they do!

In our model, objects are constantly exchanging energy and momentum via virtual photon exchange. Hence, all interacting objects move synchronously, and together, and all at the same time!

In this picture, Newtonian gravity *can be* considered as instantaneous, since celestial bodies conspire to move synchronously due to their *mutual* gravitational interaction.

In our theory, the motion of mass is the source of all interaction. This means *all* relative motion. Particles cannot tell the difference!

When you designate, and then constrain, a force center, be it a proton or the sun, you lose a small, but important contribution to the interaction and the overall motion of the system.

Fields are not real. They are mathematical models with limitations, and they are not real physical entities existing in time and space.

No fields means; no retarded potentials, no waves traveling backwards in time, and no infinite energy sums

Fields: good riddance!

The 1/R^2 law:

All matter interactions are due to real photon emission or virtual photon exchange.

Both of these manifestations of the electromagnetic interaction follow the 1/R^2 law of diminishing returns.

The mechanism for the 1/R^2 behavior is completely different for the two cases, although the reason for the behavior is exactly the same! Both are dependent on space being flat and three dimensional.

For real photons, the intensity falls off as 1/R^2 because the net flux of real particles (or energy and momentum) through a sphere of any radius surrounding the source, must be conserved.

For virtual photons, the strength of an interaction between any two particles diminishes as 1/R^2 as a consequence of the uncertainty principle.

In either case, space must be Euclidean.

Force and acceleration:

The underlying assumption of the general theory of relativity is that acceleration should not be considered a special state of motion.

The problem with this idea is that without acceleration, there is really be nothing to talk about!

Explaining acceleration (or more generally, rates of change of states) is the *raison d'etre* of physics.

Furthermore, acceleration is not relative like velocity. An accelerating object will have some acceleration in any inertial (i.e. uniformly moving) reference frame. Observers in different reference frames will compute different forces, velocities, etc., but energy and momentum will be conserved in all inertial frames.

Acceleration is *absolute motion* and is what physics must explain.

In our universal model, angular motion, such as planetary spin, is also considered acceleration and should be regarded as absolute motion. Consider two spinning bodies orbiting one another. One cannot find a reference frame where someone isn't spinning!

A model where everyone is at rest, everywhere, with no established, or prefered reference frame(s), not only removes time from the universe, but also renders the very idea of motion impossible to talk about in any meaningful way.

And yet, things move . . .

It is time to bring back motion and rates of change!

Time and space:

Time and space are conceptions of the human understanding, *and not* perceptions of the senses, representing externally existing physical realities.

There is no time or space beyond our propensity to note the rates of change of states, both conscious "states"; time, and motion; the changes in the relative positions of a system of interacting objects.

This was demonstrated by the much disparaged and maligned *thinker*, Immanuel Kant, over three hundred years ago!

Time and space are emergent properties (to use the popular parlance) of the relative motion between objects. Time and space arise from our need to assign positions and velocities to objects in our study of dynamics.

No objects; no space. No motion; no need for time!
Know objects; know space. (It had to be done!)

In any given inertial reference frame, it is appropriate to define time and space as Newtonian. We have a fixed three dimensional space and a ticking clock.

The only new feature we impose is that a particle's mass is now dependent on its velocity. However, this is *not* a comment on the nature of time and space.

Particles 'lose and gain' 'kinetic energy' during an interaction and this is reflected in the particle's varying mass-energy.

We work with the Lorentz transformations and write our four vectors in Minkowski space because a particle's mass-energy is a nonlinear function of the the velocity of the particle, *even in* our ordinary, everyday, reference frames of three dimensional Newtonian space and time. (This approach is in accord with Kant's metaphysical model of space and time.)

Now, the general theory of relativity is not just a "theory", but a *metaphysical* model of the world; a model within which people 'do physics' as conceived and prescribed by the concepts of the model.

Unfortunately, considered as either a physical theory or a metaphysical model, the general theory of relativity is incorrect.   (It's OK!)

It is incorrect about the nature of time, space, motion, and acceleration; as well as the fundamental, irreducible roles of relative and absolute motion; *and* acceleration in particle interactions.

As a metaphysical model, it is not only incorrect, but *un*-correct, as it has led researchers (and not only in the field of physics) down the wrong path.

As it turns out, time and space are 'just normal'. Time and space are just as you and I experience them and inferentially define them every day.

Time and space are 'just normal'; just as Isaac Newton envisioned.

Space just sits there and time ticks.

The choice of a particular inertial reference frame is *irrelevant* for the proper description of the interaction between particles, and this choice does not affect, or have an influence on, the intrinsic nature of time or space, or on the "behavior" of time and space.

Over the last century, the awe filled allegiance and fanatical fidelity to the general theory of relativity, has been a constant burr in the ass of progress.

The theory 'adequately' explains the bending of starlight by the sun and the perihelion of Mercury. On the other hand, the theory predicts curved and stretchy spacetime, the big bang, wormholes, singularities, dark matter, dark energy, acceleration (twice!), and many other *untenable ideas*.

Let it go. It is time.   It's alright.   :)

The Lorentz force:

It is the premise of our model that the motion of mass is the basis of all interaction and all physical phenomena. This assumption was necessary in order to afford the newly massive neutrino a magnetic moment.

For this reason, we find the charge to mass ratio, e/m_{rest}, of a *charged* particle to be the proper and fundamental *coupling* <u>constant</u> of electromagnetism (long, a popular idea, amongst many!), rather than the electric charge, e, alone, and the *relativistic* mass of the particle to be the actual *coupling* <u>charge</u>.

So, in our model, the complete relativistic Lorentz force on a particle due to a specified distribution of charge and mass is given by

$$\mathbf{F} = (me')*\mathbf{E} + (me')*\mathbf{v}\times\mathbf{B} - m*\mathbf{F\_g} - m*(\mathbf{v}\times\mathbf{B\_g}) \qquad (3)$$

where m is the relativistic mass of the particle, e' is the charge to mass ratio of the particle

$$e' = e/m_{rest} \qquad (4)$$

and m_{rest} is the rest mass of the particle.

The term F_g is the Newtonian force due to gravity and is calculated in the usual way (as tiny as it may be!)

$$F\_g = G*m1/R^2 \qquad (5)$$

where R is the distance between our particle and a gravitational source charge of relativistic mass m1.

The term B_g (very tiny!) is the 'gravitational magnetic field' vector, analogous to the electromagnetic term B, and is calculated in a similar way;

B_g = (G/c^2)*m1*(v1xR)/R^3   (6)

where v1 is the velocity of the source particle. The constant giving the strength of the force, G/c^2, is chosen to give our resulting gravitational waves the speed of light.

Equation (3) for the total force on a particle is manifestly Lorentz invariant due to the common mass factor appearing in all four terms. The electromagnetic terms reduce to the usual ('classical') expression in the low particle velocity limit, where m = m_rest.

The original motivation for 'symmetrizing' the classical Lorentz force with respect to gravity was to prop up a quite separate hypothesis, which was that the neutrino had a magnetic moment, even though the magnetic moment arises solely from the 'rotating' motion of its mass.

However, this symmetrization allows for a classical description of gravitational waves *and* bolsters our conclusion from quite separate arguments (1) concerning a gravitationally bound state of two neutrinos, that the graviton is a massless, spin 1, boson just like the photon.

In fact, we were forced to conclude that the graviton and the photon were the same particle for lack of any distinguishing characteristics! (3).

The gravitational waves would continue the electromagnetic spectrum just where the radio waves start to fade away ...

As we constructed our model, we found (or required, demanded) that the gravitational and electromagnetic interactions have the same structure, mechanism, and behavior; the only difference being the value of the electric charge.

In addition, we supposed (postulated) that the gravitational and electromagnetic interactions have the same structure, mechanism, and behavior for both macroscopic and microscopic systems; *the only difference being quantization.*

In this spirit let's have a closer look at our new Lorentz force! Let's rewrite equation (3) as

F = m*(e/m_rest)*E + m*(e/m_rest)*(vxB) + m*F_g + m*(vxB_g)   (7)

where we are not going to worry about the relative plus and minus signs between the electromagnetic and gravitational terms for the moment.

Factoring out the constant coupling charge, e/m_rest, we have

$$F = (e/m\_rest)(m*E + m*(v \times B)) + (m*F\_g + m*(v \times B\_g)) \qquad (8)$$

We see that the electromagnetic and gravitational forces now have exactly the same form and they both depend on and vary with the total relativistic mass energy of a particle.

For the electromagnetic terms, e/m_rest is *the* fundamental coupling *constant*.

The charge to rest mass ratio of a particle is now the fixed fundamental coupling constant, rather than e or alpha, and needn't be involved in calculations except as a multiplicative factor.

We note, when m = m_rest, the electromagnetic and gravitational forces are totally 'decoupled'.

Now, let's take a closer look at the new mass dependence of the Lorentz force as described by equation (8). For simplicity, we will focus only on the two electromagnetic terms. The force on a massive charged particle (e.g. an electron) due to a specified distribution of mass and charge is then

$$F = (e/m_e)(m*E + m*(v \times B)) \qquad (9)$$

If we consider the interaction between two identical electrons, we must assume one provides the field for the other, and equation (9) becomes

$$F = (e/m_e)(m_1*E + m_1*(v_1 \times B)) \qquad (10)$$

Using the well known equations for **E** and **B** due to the second electron, we have

$$F = (e/m_e)^2(1/R^2)(m_1 m_2 (1/4\pi\varepsilon)r + p_1 \times p_2 \times r(\mu/4\pi)) \qquad (11)$$

Some factoring yields

$$F = (e/m_e)^2(c/R)^2(\mu/4\pi)(m_1 m_2 r + (1/c^2)(p_1 \times p_2 \times r)) \qquad (12)$$

For non-relativistic interactions, we can solve the relationship

$$E = p^2/2m \qquad (13)$$

for the mass of our two electrons, finally yielding

$$F_{1,2} = (e/m_e)^2(c/R)^2(\mu/4\pi)(p_1^2 p_2^2/4E_1 E_2 + (1/c^2)p_1 p_2 \sin\theta) \qquad (14)$$

assuming our two electrons are moving parallel to one another.

The purpose of this exercise was to demonstrate that the classical Lorentz force between two particles can be formulated solely in terms of their energy and momentum.

The classical coupling strength is now given by our new universal coupling parameter, $(e/m\_e)^2$, which can be carried over unscathed and unmolested into quantum mechanical calculations!

We now reintroduce the gravitational terms into equation (12) and find the the complete, 'classical', relativistic, Lorentz force between two identical electrons

$$\mathbf{F} = (G/c^2 - (e/m_e)^2(\mu/4\pi))(c/R)^2(m_1 m_2 \mathbf{r} + (1/c^2)(\mathbf{p_1} \times \mathbf{p_2} \times \mathbf{r})) \qquad (15)$$

Equation (15) demonstrates the *equivalence* of inertial and gravitational mass.

We note that the factor of $(1/c)^2$ dampens the gravitational interaction relative to the electromagnetic interaction. In addition, both the gravitational and electromagnetic 'magnetic' terms are dampened by a factor of $(1/c)^2$ relative to their respective static forces.

Also, notice that the factor of epsilon_0 is 'gone', and we are left with two coupling constants, G and e/m_rest, and a 'space factor', $\mu/4\pi$. The evolution of the force is totally described by the energy and momentum of the two electrons. The factor $(c/R)^2$ describes the time dependence of the interaction, and the time t is *common* to both electrons.

The particles act on one another. There is no 'source' charge.
This description contains no electric or magnetic fields!
All the energy and momentum of the system is carried by the particles.

Nowadays, we characterize this interaction as two electrons trading energy and momentum by virtual photon exchange. Exchange is the key word. There is no emitter and no receiver; just a continual flux of virtual photons between the two.

Actually, a picture I like better, is of one of a continuous virtual photon, constantly changing 'frequency' or 'mass' as the interaction evolves.

Let's look again at the force between two identical electrons as described by equation (15).

We note, of course, that $\mathbf{F_1} = -\mathbf{F_2}$. Newton's second law tells us

$$\mathbf{F_1} = m_1 \mathbf{a_1} \qquad (16)$$

If we compare equations (15) and (16) we can see that the acceleration of a particle is *independent of its mass*.

$$\mathbf{a_1} = \text{Function}(m_2, \mathbf{R}) \qquad (17)$$

This is a general result that we used to assume applied only to the gravitational interaction (and is used as an argument in favor of general relativity).

In addition, the relative strengths of the four terms in equation (15), or the 'four forces of classical physics', are approximately as follows;

electricity = 1 ; magnetism ~ 1/c^2 ; gravity ~ G/c^2 ; magnetic gravity ~ G/c^4

I think Maxwell would approve!; except that we have no more need for his fields.

The two electrons exert equal and opposite forces on one another during the interaction, and the evolution of the force is completely described by the (variable) mass-energy of the two electrons, $m_1(\mathbf{r_1}(t))$, $m_2(\mathbf{r_2}(t))$, where the time, t, is *common* to both electrons.

Of course, without the fields there is no mathematical or physical mechanism to explain the interaction of these two particles 'at a distance'.

We like the idea of 'one virtual photon' (which, of course, is a discovery of the field theory model) constantly coupling the particles; a time varying conduit for energy and momentum exchange. Can we infer or derive the virtual photon without resort to field theory?

We will defer this investigation for now.

Let's look at our 'new' Lorentz force in a little more detail. If we define

$$K == (G/c^2 - (e/m_e)^2 (\mu/4\pi)) \qquad (18)$$

then we can write equation (15) as

$$\mathbf{F} = K*(c/R)^2 (m_1 m_2 \mathbf{r} + (1/c^2)(\mathbf{p_1} \mathbf{x} \mathbf{p_2} \mathbf{x} \mathbf{r})) \qquad (19)$$

Next, we factor out the particle masses from the momentum term

$$\mathbf{F} = K*(c/R)^2(m_1 m_2 \mathbf{r} + (1/c^2)(m_1 m_2 \mathbf{v_1} \mathbf{x v_2 x r})) \tag{20}$$

$$\mathbf{F} = K*(c/R)^2(m_1 m_2)(\mathbf{r} + (1/c^2)(\mathbf{v_1 x v_2 x r})) \tag{21}$$

where **r** is the unit vector **R**/R.

Since our two masses form a closed, conservative system, we can 'normalize' our force by dividing by the total energy of the system; $E_{TOT} = m_1 + m_2$.

$$\mathbf{F}/E_{TOT} = K*(c/R)^2 \mu (\mathbf{r} + (1/c^2)(\mathbf{v_1 x v_2 x r})) \tag{22}$$

where $\mu(\mathbf{R}, d\mathbf{R}/dt) = m_1 m_2/(m_1 + m_2)$ is the reduced mass of the two body system.

In order to investigate the vector cross product term, we will assume our two masses (no longer necessarily electrons) are equal and orbiting one another.

Then we can write

$$\mathbf{F}/E_{TOT} = K*(c/R)^2 \mu (1 - (v^2/c^2)) \tag{23}$$

Seriously! And

$$\mathbf{F}/E_{TOT} = K*(c/R)^2 \mu - K*(\mu v^2/R^2) \tag{24}$$

We will call the second term in equation (24), the *coriolis* force.

If we recast our force equation into polar coordinates and allow $m_1 \neq m_2$ (i.e. for the study of planetary motion; Kepler's equations), we will pick up the usual *centrifugal* force term, in addition to our new *coriolis* force term.

$$\mathbf{F}/E_{TOT} = K*(c/R)^2 \mu - K*(\mu v^2/R^2) - K*(l^2/\mu R^3) \tag{25}$$

Finally, there will be a force term corresponding to the interaction of the spin/angular momentum (σ, l) of one object with the 'magnetic force vector' of the other object; F_{spin}.

(Remember, in our model, the relativistic mass of an object is due to the *total* relativistic motion of the mass; including spin!)

So, the total Lorentz force, for cosmology for example, will consist of four terms;

$$F_{universal} \sim F_{central} + F_{coriolis} + F_{centrifugal} + F_{spin} \qquad (26)$$

In our generalized Lorentz force, the coriolis force is not an artifact of the choice of a particular reference frame, but arises from the absolute relative motion of the two bodies.

In our model, the coriolis force is *real*, because all forces are velocity dependent due to the fact that particle mass is velocity dependent; $m = m(\mathbf{r},\mathbf{v})$.

When we speak of the relativistic mass in our model, this includes *all relative velocities*, and not only linear velocities approaching the speed of light.

Similarly, the centrifugal force is now also a real velocity dependent force and is not due to "space time" disturbances as in the general theory of relativity.

Equation (25) should be 'easy' to generalize to an N body system.

The Lorentz torque:

The 'Lorentz torque' arises from the interaction of the intrinsic angular momentum (quantum spin or classical moment of inertia) of one body with the magnetic field vector of the second body, yielding the force 'F_{SPIN}' introduced above.

Unfortunately, this 'force' is actually a torque, which seems to complicate adding it to our Lorentz force equation. However, if we consider the potential energy instead of forces, the problem goes away!

$$E_{SPIN}/(E_{TOT}) = (\mathbf{I_1} \cdot \mathbf{B_2})/(m1+m2) + (\mathbf{I_2} \cdot \mathbf{B_1})/(m1+m2)$$

$$E_{universal} \sim E_{central} + E_{coriolis} + E_{centrifugal} + E_{spin}$$

Cosmology:

The general theory of relativity is no longer thought to describe the gravitational interaction of matter. Hence, many of the current problems in cosmology, which imply the inadequacy of the current theory, need to be reevaluated in the light of our new theory.

Large scale neutral matter interactions are described by the cosmological Lorentz force

$$F = m^* F\_g + m^* (v \times B\_g) \quad\quad\quad (27)$$

It is important to note, that the mass of the objects involved is the *total relativistic mass*, which includes the mass due to rotation.

Angular motion is acceleration. It is also translationally invariant and thus considered to be absolute motion. Thus when considering the interaction between two galaxies, for example, one must include the relativistic masses of the individual spinning bodies, as well as the contribution to the total relativistic mass of the galaxy due to the bodies rotating about the center of the galaxy!

A body on the edge of a rotating galaxy will have a relativistic mass relative to the center, modifying the central force on the body and its angular velocity.

No more dark matter!

We can also now see why planets bulge. Matter on the equator of a planet is more massive than matter at the poles.

Also, there is now the prospect for repulsion between two bodies moving in opposite directions. In addition, a spinning galaxy now creates a 'gravitational magnetic field' and can be assigned a series of magnetic moments.

Obviously, if equation (27) is correct, then all the derivations already done in the study of electrodynamics can be applied directly to cosmology!

Our (*independently*) derived) theory is very similar to Gravitoelectromagnetism (GEM).

The only thing the GEM equations (analogs of the Maxwell equations) are missing is the mass current term, **J_m**, in the equation for electromagnetic induction, which we introduced in "On Parity and Isospin".

Red shift:

The gravitational contribution to the redshift of light from a distant galaxy is no longer attributed to the stretching of space. Instead, the shift in frequency is due to the gravitational work done on the photon by the mass of the emitting galaxy. This view will yield new relative velocities for the various galaxies, as well as, perhaps, new mass estimates.

Black holes:

In our new model, ordinary matter is composed of electrons, positrons and antineutrinos.

When the mass of a body reaches some critical point and the pressure becomes great enough that the force of gravity overwhelms the electrostatic repulsion responsible for the structure of the protons and neutrons, the constituent electrons and positrons will annihilate producing photons. Of course, we will also have the residual electron antineutrinos.

So, we would have a mix of bosons and fermions. It seems the photons would all fall into the same lowest possible frequency energy state, perhaps defined by the diameter of the black hole, and form a Bose-Einstein condensate. The antineutrinos might either escape, or remain gravitationally trapped, thus giving the black hole it's physical extent (and extra mass) due to the 'Pauli exclusion principle' keeping the neutrinos apart.

In our model, black holes consist of photons and electron antineutrinos, as well as whatever other detritus happens to get sucked in.

A black hole can be considered to be an 'infinite' spherical potential well of radius R. We imagine that the gravitational potential is constant inside the black hole

$$V(r) = G*M/R \; ; \qquad ; r < R \qquad (28)$$

and falls off in the usual way for r > R,

$$V(r) = G*M/(R+r) \qquad ; r > R \qquad (29)$$

The energy levels of the antineutrinos inside can be calculated using well known techniques. Since the potential energy of the well is not actually infinite, there will be the possibility for the high energy neutrinos to tunnel their way out!

For the photons, we don't like the idea of 'plane wave' solutions (as they would be reflecting off of a "potential barrier"), and so propose instead, a collection of di-photon bound states.

It's all quite speculative at the moment!

The eternal universe:

We predict the picture of the universe that will emerge after a reevaluation of the current cosmological data, will be one of an eternal, infinite, and *static* universe.

From lowly elementary particle decays to exploding stars, our universe is constantly being reseeded with new materials to build new stars, new solar systems, new galaxies.

On time and temperature:

Entropy.

All together now …   boring and confusing!

Fortunately, there is no room for entropy in our cosmological model.

There is no equation of the universe. Nothing so grandiose!

Our universe is an infinite, static, eternal, yet 'closed system'.

Our model of the universe can be considered as 'closed' in two ways; 1) in the sense that there are no "external" influences!, and 2) any system of interest (electron-proton, earth-moon, earth-sun, galaxy-galaxy, etc.) *can be*, and are routinely, appropriately and effectively isolated, both experimentally and theoretically, to produce sensible predictions and results.

For cosmological studies, one would choose an appropriately 'isolated' system (e.g. all the galaxies in the local cluster) and apply the analogues of equations (25) and (26).

Space is space. Space is the same here, there, and everywhere. From the space between atoms to the space between the stars,

It is all the same space.

Quantum mechanical interactions:

In our new model, the photon facilitates all interactions between particles, just as in quantum electrodynamics, except that in the universal model the photon couples to the relativistic mass of a particle as well as the electric charge.

We shall construct our new quantum mechanical coupling charge in strict analogy with the classical coupling charge of equation (15).

The photon then couples to a particle's 'total coupling charge', tcc, which is defined (with the sign convention fixed by equation (15), rightly or wrongly!) to be

$$tcc = (m/a)*(b - e') \tag{30}$$

where m is the particle's relativistic mass and e' is the charge to mass ratio as before, and

$$a = (4*pi*epsilon\_0*hbar*c)^{1/2} \tag{31}$$

and

$$b = (4*pi*epsilon\_0*G)^{1/2} \tag{32}$$

The tcc is designed to yield

$$alpha = e^2/(4*pi*epsilon\_0*hbar*c) \tag{33}$$

for purely electromagnetic (and non-relativistic) interactions, and

$$alpha\_G == m\_e^2*G/(hbar*c) \tag{34}$$

for gravitational interactions. (There will be an 'unfortunate' cross term between alpha and alpha_G in charged particle interactions, which we will discuss later on.) Anyway ...

$$tcc = (m)(G^{1/2} - (e/m_e)(1/4\pi\varepsilon)^{1/2})(1/h^{bar}c)^{1/2} \tag{35}$$

For quantum mechanical calculations, the propagator for the interaction between any two particles is then

$$f(q) = (tcc\_1)*(tcc\_2)/q^2 \tag{36}$$

where q is the four momentum of the exchanged photon.

We can see from equations (35) and (36), that the energy scale dependence of an interaction is now determined by the mass of the interacting particles, rather than by the mass of the propagator as described in the standard model. In our new model, the propagator is always massless.

Finally, the photon also has a coupling charge, m_g, given by

$$m_g = \hbar \nu / c^2 \qquad (37)$$

where ν is the frequency of the photon and c is the speed of light.

Nonrelativistic quantum mechanical gravitational interactions:

We began our enquiries in reference (1) by considering a gravitationally bound state of two identical electron neutrinos; neutrinium. We assumed our two neutrinos would be bound by the classical Newtonian gravitational potential and thus based our neutrinium model on the analogous, and well known, positronium system bound by the Coulomb potential.

The Bohr radius was derived (missing an all important factor of G, here remedied!) to be

$$R_{neutrinium} = 2(4\pi \hbar^2)/(G m^3) \qquad (38)$$

where m is the neutrino mass. We then compared this result to that expected from general relativity,

$$R = 2 h^2 / (G m) \qquad (39)$$

Noting the discrepancy of a factor of $1/m^2$ between the two results, we had to conclude that the general theory of relativity was incorrect (1,2).

It had a good run!

The strong force:

We remind the reader that we have defined the gravitational "rest charge" of a particle to be equivalent to its rest mass.

We boldly extend our gravitational model to the strong force and assume that a particle has a "rest color charge" analogous to our gravitational rest mass charge. A particle's color charge would then increase with its momentum. This model would account for quark confinement and asymptotic freedom.

At this point, it actually seems unnecessary to keep the color charge at all, so we declare

rest color charge == rest mass charge

If this is the case, then the strong force is just relativistic gravity.

Since we have no more need for the color charge, it seems reasonable to dismiss the quarks as well, and assume all partons are actually leptons.

In this model we have,

proton   = (e+,e-,e+)

neutron = (e+,e-,$\bar{v_e}$)

electron =   e-

and there is no more mystery of the missing antimatter. Bonus!

In the universal model, the strong force is merely the relativistic gravitational interaction balanced by the electromagnetic interaction, plus an additional interaction due to our surprising new 'cross-term'!

In our new strong force model, the linear force term arises from the mass of a particle and is multiplicative; rather than including an additional force term which is linear in the separation R.

Also, there is a natural attenuation of this force as particles approach relativistic velocities; hence, it can never become infinite.

As an added bonus, there are no more fractionally charged quarks, summing over colors, etc.

Let's look at the total coupling between two electrons.

For the sake of simplicity, and for illustrative purposes, we will begin with the non-relativistic case, where m*e' ==> e.

The total coupling, $(tcc\_e)^2$, is then, from equation (30)

$$(tcc\_e)^2 = (1/a^2)*(e^2 - 2*e*m*b + m^2*b^2) \tag{40}$$

Using equations (31) - (34), we have

$$(tcc\_e)^2 = alpha - 2*e*m*b/a^2 + alpha\_G \tag{41}$$

We shall call the cross term, alpha_strong, where

$$alpha\_strong == (G/(4*pi*epsilon\_0))^{1/2}*(2*e*m/hbar*c) \tag{42}$$

Remembering that equation (42) is non relativistic, we can replace e with m*e' to obtain

$$alpha\_STRONG = (G/(4*pi*epsilon\_0))^{1/2}*(2*e'*m^2/hbar*c) \tag{43}$$

We see from equation (41) that our new strong force term acts as a counterbalancing force, changing sign for different combinations of electrons and positrons.

The coupling alpha_strong would be appropriate for describing interactions between nucleons (protons and neutrons), while alpha_STRONG would be the choice for parton interactions within the nucleons.

In our model of the proton and neutron (and all baryons), the leptons are highly relativistic, so no simple hand waving model is possible. On the other hand, we do not have to worry about, or demand, antisymmetric parton wave functions even for identical partons within a baryon.

The running of alpha:

In the standard model, the electromagnetic coupling strength is expressed in terms of the electric charge, e;

$$\alpha = e^2/4\pi\varepsilon \hbar c \qquad (a)$$

In our model, the 'electric charge' looks like this

$$e \rightarrow (m^*e/m_e) \qquad (b)$$

and alpha becomes

$$\alpha = m^2(e/m_e)^2/4\pi\varepsilon \hbar c \qquad (c)$$

$$\alpha = (m_e^2/1 - v^2/c^2)(e/m_e)^2/4\pi\varepsilon \hbar c \qquad (d)$$

$$\alpha = (1/1 - v^2/c^2) e^2/4\pi\varepsilon \hbar c \qquad (e)$$

$$\alpha = \alpha_0 (1 + (v/c)^2 + (v/c)^4 + \ldots) \qquad (f)$$

In our running of alpha, the velocity squared replaces the four-momentum transfer, Q^2, of the standard model and there is no need to introduce an arbitrary cut-off mass.

In the standard model, the higher order Feynman diagrams for a scattering process represent the perturbative expansion of a *single integral*, in terms of the fixed coupling constant alpha, leading to the *running* of alpha, charge screening, renormalization, etc.

These higher order diagrams represent vacuum polarization loops, multiple photon exchanges, and all the other pathologies of quantum field theory.

In our model, there *is* only one integral as there *is only one* virtual photon involved in the interaction or scattering. There are no fields, there is no vacuum, and there is no charge screening.

In our model, e/m_rest is a fixed coupling constant and the relativistic mass of a particle is the only variable, so our perturbative expansion of alpha is in terms of $(v/c)^2$.

Rather than imagining a "high Q^2 probe" penetrating a virtual electron cloud and thus increasing the effective coupling charge, we have an increased coupling charge because the particles involved in the scatter are highly relativistic.

The weak interaction:

In the universal model, weak interactions always involve the neutrino, and thus are weak due to the tiny neutrino mass.

One of the conclusions of "A Quantum Mechanical Theory of Gravitational Interactions" was that a particle's 'weak charge' was equivalent to its mass, and that this tiny coupling charge was enough to explain the 'weakness' of the weak interaction.

Even so, we did not address the existence of the W and Z bosons out of convenience and expediency (and with no clear ideas on the matter at the time!). In this section, we will examine whether the W and Z are really necessary for weak interactions, and whether we can explain them away!

Let's begin historically, with beta decay; whereby a neutron turns into a proton;

$$n \longrightarrow p + e- + v\_e\textasciicircum bar \qquad (44)$$

In the standard model, beta decay occurs when a down quark emits a massive W- boson to become an up quark. The W- particle then decays into an electron and an electron antineutrino.

$$(u,d,d) \longrightarrow (u,u,d) + e- + v\_e\textasciicircum bar \qquad (45)$$

This process is highly suppressed due the massiveness of the W and not due to the nature of the weak charge.

In our new model, where partons are leptons, beta decay now looks like this

$$(e+,e-,v\_e\textasciicircum bar) \longrightarrow (e+,e-,e+) + e- + v\_e\textasciicircum bar \qquad (46)$$

Here, the antineutrino emits a photon which then decays into an electron and a positron. In terms of the Feynman diagram, the coupling charge at the antineutrino vertex is the neutrino mass (which 'suppresses' the interaction).

$$\text{alpha\_weak} == m\_v^2 * G / (hbar * c) \qquad (47)$$

The coupling charge at the electron vertex is the tcc of the electron.

This view of the weak interaction does not require a charged quantum mechanical force field. It is already a 'stretch' to imagine fields possessing energy and momentum, much less mass and electric charge!

As a second example, let's look at muon decay,

mu+ ---> v_mu + e+ + v_e (48)

Again, in the standard model, the muon emits a charged W particle turning into a muon neutrino. The W+ then decays into a positron and an electron neutrino.

In addition to requiring a virtual field to carry the mass and electric charge away (as in beta decay), it just seems unnatural that a particle should change into another particle, particularly in such an *ad hoc* way.

For muon decay, we imagine the muon emitting a muon neutrino just as in the standard model, but rather than coupling to a virtual W, the muon couples to a virtual, massless, charged lepton which then emits an electron antineutrino, thus becoming a real electron.

This process is shown in Figure 1.

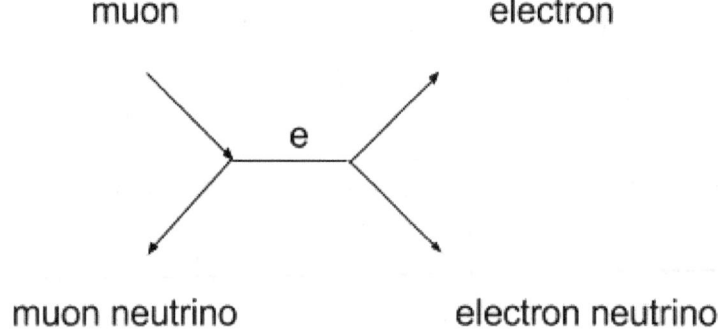

FIGURE 1: Muon decay. The propagator is most likely a generic virtual lepton, e/mu/tau.

In our model, the muon 'sheds' energy and spin in the form of its neutrino. The energy and spin shed ensures that the ensuing virtual lepton propagator has spin = 0, and is *massless*. In our model, all 'propagators' are massless.

This requirement places constraints on the energy and momentum of the initial and final states in muon (and tau) decays, and should help to explain the energy hierarchy of the three particle families. (A homework assignment!)

So, we have accounted for the weak interaction, with a standard model type approach, but without having to invoke charged, massive virtual exchange bosons or fields.

This new weak theory is essentially QED with mass as the coupling charge. Beauty, eh?

"Hold on", you might say, "haven't the W and Z been observed in particle collisions?"

In our new theory, these particles represent temporary resonant bound gravitational states of the colliding leptons. (For W production in ppbar collisions, it is equally likely to imagine the event as deep inelastic scattering between an electron and a neutrino.)

Our conclusion is the electroweak theory gives the correct energy scale for the unification of electromagnetism and the 'weak interaction', but an incorrect picture of the mechanism.

As an example, let's consider Z production in electron positron collisions.

In the standard model, the electron and positron annihilate producing a Z (or technically Z/gamma!) which can then decay into a pair of leptons or a pair of quarks which then decay into 'jets' of particles. So, our new model must explain this hadronization.

In the universal model, the electron and positron can annihilate to a photon or form a temporary, gravitationally bound, mesonic state.

Once the electron and positron are bound gravitationally, gravity effectively starts to behave like QCD in terms of confinement. The particles can only 'break apart' by forming gravitational bonds with neighboring particles.

Of course, this is only a rough analogy. We will need a slightly more rigorous picture of the gravitational hadronization and the gravitational strong forces for a complete model.

More homework!

The electroweak bosons:

The massive bosons of the standard electroweak theory are no longer thought to be 'field quanta', but are now considered to be resonant bound states of the two fundamental leptons.

$W^+ = (e^+, \nu_e)$

$W^- = (e^-, \bar{\nu_e})$

$Z = (e^+, e^-)$

The Z looks to be an (higher) excited state of positronium; the first excited state would be assigned to the pion. Or perhaps, the Z could be made up muons instead, or an admixture of all three (known) charged leptons!

In this model, the newly observed Higgs' boson would then be a bound state of an electron neutrino and an electron antineutrino.

$H = (\nu_e, \bar{\nu_e})$

The vacuum:

In the universal model, the Higgs boson is no longer required to provide the particles with mass, and indeed, the universe is no longer permeated by the Higgs field (or the W field, or the Z field, or the quark and gluon fields, etc.!).

In the standard model, the Higgs boson was thought to arise from a symmetry breaking of the energy of the vacuum. Now, this apparent feature of the 'physical vacuum' no longer seems to be necessary.

In our new model, the vacuum has no energy and no symmetry (or asymmetry for that matter!) and is not a teeming cauldron of subatomic particles.

In our model, fields are not real and hence there is no quantum mechanical vacuum.

Regardless, even in quantum field theory, when performing the 'second quantization' of particle fields in terms of the simple harmonic oscillator, one has the option of retaining the *constants of integration* (thus *creating* the quantum vacuum!), *or* setting them to zero.

Such a matter of taste should not be the basis for any model of reality.

The Dirac equation:

Why is the normalization for a Dirac spinor $1/(E + m)^{1/2}$?

This is quite unsettling since the normalization is greater than the total energy of the particle by an additional amount corresponding to the rest mass of the particle.

This cannot be right!

The Dirac equation for a 'free electron' is

$$H \psi = (\alpha \cdot p + \beta m_0) \psi \qquad (49)$$

where the Hamiltonian H, is the total energy of the electron, and m_0 is the rest mass. The Hamiltonian operator is

$$H = i \, \partial/\partial t \qquad (50)$$

In our model, the mass-energy, $E = mc^2$, is the particle's coupling charge, and completely determines the particle's behavior. So the rest mass term in the Dirac equation seems, somehow, 'redundant', unnecessary, and *unwanted*.

Let's define the kinetic energy operator (T = m - m_0) to be

$$T = H - \beta m_0 \qquad (51)$$

We still don't like, or want, this extra, constant, additive, and *global* charge mucking up our beautiful equation, so (and I think you see what's coming ...) we make a '*global gauge transformation*' and the problem is gone!

The Dirac equation for a *massive* electron is now

$$T == H \psi = (\alpha \cdot p) \psi \qquad (52)$$

and the spinors for the left handed and right handed solutions are now *decoupled*.

In our model, the Dirac equation does not have positive and negative energy solutions. (A 'free particle' cannot have negative energy!) Instead, we have spin left (electrons) and spin right (positrons) solutions. There is no right handed electron in our model or in nature!

There is no positron sea, and no particles traveling backwards in time.

If we add interactions to the mix, the general form of the Dirac equation will be

$$(T - V) \psi == L \psi = 0 \qquad (53)$$

where, of course, L is the Lagrangian operator.

Our 'gauge invariant' solution of the free particle Dirac equation would look something like;

$$\psi = \exp(-i^*m_0) \exp(-iE^*t) \exp(i\mathbf{p} \cdot \mathbf{x}) \qquad (54)$$

Replacing E(t) with the relativistic mass m(t), gives

$$\psi = \exp(-i^*m_0) \exp(-im^*t) \exp(i\mathbf{p} \cdot \mathbf{x}) \qquad (55)$$

The first factor in equation (55) is a global phase factor representing the invariant rest mass-energy/charge of the particle. This is an extra and annoying charge which is already included in the second factor of the equation; i.e the total mass-energy/charge, m(t), which we take to be our locally variable 'phase factor' *and* the locally variable interaction charge of the particle.

However, our theory is not really a gauge theory, and there seems to no formal way (e.g. by a series expansion) to have the global phase factor in equation (55) cancel the rest mass energy term in the Dirac equation.

So ... we take our lead from the standard model and throw the rest mass term away!

However, we needn't 'recover' the rest mass from an interaction with the Higgs field as in the standard model. In our model the contribution of the rest mass energy is already accounted for in the total relativistic mass energy of the particle, which is the particle's interaction charge.

To introduce interactions into the Dirac equation, we modify the standard replacement

$$p_\mu \to p_\mu + eA_\mu \qquad (56)$$

to include the gravitational interaction, and to account for the relativistic mass as the fundamental interaction charge, and obtain

$$p_\mu \to p_\mu - i^*\hbar (G^{1/2} - e/m_{rest}) A_\mu \, \partial/\partial t \qquad (57)$$

This substitution should then yield the Universal Interaction Lagrangian

$$L_{interaction} = -i\hbar(G^{1/2} - e/m_{rest})(\psi^{bar}\gamma^{\mu} A_{\mu} \partial\psi/\partial t) \qquad (58)$$

We interpret the component A_{μ} as representing a *virtual* photon, and thus we have no need for the antisymmetric tensor $F^{\mu\nu}$, and the corresponding E and B fields, to facilitate the interaction or to carry energy and momentum. Remember, in our model, all the energy and momentum in an interaction is carried by the particles!

The virtual photon is already coupled to, or 'tethered' between, two mass-charge currents, only one of which is indicated in equation (58). The second current could be a similar leptonic current or a real photon current!

The wave function of the virtual photon propagator would be

$$A_{\mu} = \varepsilon(\mu) \exp(-i((m_1-m_2)/\hbar) t) \qquad (59)$$

where m_1 and m_2 are the relativistic masses of the interacting particles.

In our model, we expect to treat the real and virtual photons as separate particles, each with their own 'wave function'.

Gauge theory:

We can now understand the origin of gauge invariance in QED.

The variable mass charge involved in an interaction is compensated for by a time dependent 'phase factor' in the wave function; $\exp(-iE*t)$.

Technically, our theory is *not* a gauge theory, although it does exploit and explain the observed gauge invariance of the electromagnetic interaction (another ingenuous discovery of the standard model, to be sure!).

In our model, the resolution of the 'gauge invariance issue' does not involve, or allow, the capricious variation of local charge with corresponding and compensating potential fields. (This is a feature never truly exploited, or properly explained, or *necessary* in QED ....)

Instead, in the universal model, we have a locally varying charge because the mass-energy is the coupling charge of a particle, and this varies during an interaction.

In conclusion, since the standard model explanation (and implementation) of the required 'gauge invariance' observed in QED is now thought to be incorrect, the generalization of gauge theory to explain the weak and strong interactions (thus yielding the W, the Z, and eight gluons) is probably also incorrect.

Of course, we have already removed the W, the Z, and gluons and quarks from our model. However, the more corroborating arguments for a new theory, the better.

Gauge invariance:

The universal model takes advantage of, and explains, both global and local gauge invariance.

Global gauge invariance allows us to remove the rest mass of a particle from our equations, because it is a global, constant, and therefore uninteresting, contribution to the particle's variable mass-energy charge.

Local gauge invariance is necessary because a particle's mass-energy charge varies during an interaction.

The Born approximation:

In the Born approximation, one assumes the scattering interaction between two particles can be described as the exchange of one virtual photon. This photon is considered to be the most energetic of all the photons exchanged during the interaction, and the contributions from "higher order" photons are found to be negligible in cross section calculations.

One of the mysteries of the standard model is why this first order approximation from the perturbative expansion of the interaction in terms of the coupling constant alpha, yields an exact result.

In our model, *there is only one* virtual photon involved in the interaction between the two particles, and its behavior is described by equation (59).

Hard scattering occurs as the virtual photon exchanged between the two particles becomes highly energetic. The frequency of a "hard scatter" in an experiment depends on the incident particle beam densities and individual particle impact parameters (and not on some random, or probabilistic, emission of a high Q^2 virtual photon from the 'target'!)

In the single particle exchange picture of the standard model, one particle is the emitter, and the other is the receiver. This is an excellent mathematical representation. Physically, it is lacking as the exchange is not 'symmetric', and it is hard to understand how such an exchange could lead to an attractive force.

In reality, the two particles have been interacting and exchanging virtual photons long before they were even collimated into colliding beams! However, we needn't go that far back in time.

The point is, in their final approach down the straightaway, the two particles are interacting furiously. The particles are always interacting; before, during, and after the 'scatter'.

Energy and momentum are always being exchanged between the two particles. There is no emitter and no receiver.

In our model, the q^2 value of the 'single exchanged photon' represents the total q^2 value given by the integral of equation (59).

For every action there is an equal and opposite reaction.

The classical propagator:

We now revisit our idea of creating a 'propagator' from our new universal Lorentz force.

The complete, 'classical', relativistic, Lorentz force between two identical electrons is

$$\mathbf{F} = (G/c^2 - (e/m_e)^2(\mu/4\pi))(c/R)^2(m_1 m_2 \mathbf{r} + (1/c^2)(\mathbf{p}_1 \times \mathbf{p}_2 \times \mathbf{r}))$$

where, of course, $\mathbf{F}_1 = -\mathbf{F}_2$.

We shall start our study slowly, and consider only the static, or Coulomb potential, as one usually does.

$$\mathbf{F}_{1,2} = K^*(c^2/(\mathbf{r}_1 - \mathbf{r}_2)^2 (m_1(\mathbf{r}_1, \mathbf{v}_1))(m_2(\mathbf{r}_2, \mathbf{v}_2))$$

The intriguing factor of c^2/R^2 (the result of some factoring), already looks 'propagator like', with units of $1/t^2$.

We could look at the work

$$W_{1,2} = -W_{2,1} \Rightarrow W_{1,2} + W_{2,1} = 0$$

or even the action. I believe the key to this problem is converting the integrals such that we are integrating over the masses, dm, of the two particles; the limits of integration being the initial and final relativistic masses of the two particles.

We expect the solution will look a lot like equation (59)

The building blocks of matter:

In the universal model there are two fundamental fermions, the electron and the electron neutrino; and one fundamental boson, the photon. From these building blocks, all ordinary matter is formed

The electron neutrino is thought to be the fundamental unit of mass. It is considered to be a 'true' point particle of spin ½. From a consideration of the relative strengths of the electromagnetic and weak interactions, the rest mass of the neutrino has previously been derived (1) to be

$$m\_v = m\_e/e \qquad (60)$$

where m_e is the electron mass and e is the magnitude of the electric charge.

The electron neutrino is also believed, from arguments of beauty and symmetry (13), to have a magnetic moment given by

$$mu\_v = (m)(hbar/2*m\_v) \qquad (61)$$

where m is the relativistic mass and m_v is the rest mass of the neutrino.

In our model, the motion of mass is considered to be the ultimate cause of all magnetic and electrical forces or fields. The 'spinning mass' of the (point) neutrino gives rise to its magnetic moment.

In addition, the electron neutrino is thought to 'spin to the left', while the antineutrino spins to the 'right'.

In the universal model, leptons now have two spin degrees of freedom! (?)

All leptons have a 'fixed spin' of left or right, as well as an 'interaction spin' which can point up or down. Particles spin left and anti-particles spin right. Particles maintain their defining direction of spin (left or right), while the spin ½ component can flip up or down during an interaction.

So, in our model the neutrino has an antiparticle. Neutrinos spin to the left and antineutrinos spin to the right. Similarly, electrons spin to the left and positrons spin to the right.

This explains why we only observe left handed neutrinos in interactions with the electron.

Matter is left handed. Antimatter is right handed. We live in a sinister universe!

Since we consider the electron neutrino to be the fundamental unit of mass, we shall now invert equation (1) to obtain a formula for the electron mass in terms of the neutrino mass and the value of the electric charge

$$m\_e = e*m\_v \qquad (62)$$

The neutrino mass can be taken from measurement or, ideally, derived from first principles!

We hypothesize that the electron neutrino is the "self-gravitationally" bound state of 'one quantum of action', h. We will revisit this idea in a later section.

We can now construct the universal model of the electron magnetic moment in strict analogy with the neutrino magnetic moment of equation (61), and in accordance with our new generalized Lorentz force, and the nature of our new coupling charge.

$$mu\_e = (m)(e/m\_e)(hbar/2) \qquad (63)$$

using equation (62)

$$mu\_e = (m)(1/m\_v)(hbar/2) \qquad (64)$$

and equation (61)

$$mu\_e = (1/m\_v)(mu\_v) \qquad (65)$$

In this model, according to equations (62) and (63), the electron is 'just' the neutrino with an electric charge. The electron has increased in mass-energy and coupling charge by a factor of e relative to the neutrino.

We hypothesize that the electron is an 'excited' neutrino and, that the electron's magnetic moment *and* its electric charge are caused by the motion of spinning mass.

We have called this idea 'quantum mechanical electromagnetic induction'. Here, opposite directions of spin would generate opposite electric charges (i.e. either electrons or positrons). The generated electromotive force (Joules/Coulomb) of the spinning mass would manifest itself as an electric point charge.

We assume the rest mass and the spin of the electron are constant, so this would imply that electromotive force is quantized like all other physical quantities and processes.

Also, the neutrino would have no charge in this model because it is considered a true point particle without 'a middle'.

In our new model, the electron is 'not quite' a point particle, although it is considered to be an electric point charge! The (extended) rotating mass of the electron is thought to give rise to the point electric charge by "quantum mechanical electromagnetic induction" (3). This electrical energy seems to contribute to the electron rest mass, and the value of e seems to be tied to the quantization of the mass of charged particles as demonstrated by equation (62).

Similar relations are assumed to hold for the muon and the muon neutrino as well as for the tau and the tau neutrino. Hence, we can easily predict the masses of the muon neutrino and the tau neutrino from the measured masses of the muon and the tau.

The particle model:

Particles are not waves, or wave packets, or fields, or the disturbance of fields. Particles do not pop out of the vacuum. Particles are not wave functions. A particle is not a Dirac delta function. Particles do not interact with fields, either to exert or experience force, nor to acquire mass. Fields are not real.

These are mathematical models of particles; descriptions of particles; descriptions of the behaviour of particles; descriptions of the interaction of particles

They are not particles. They are not even an *approximation* of particles.

Nowadays, it has become the fashion to state that all particles are actually 'field-particle hybrids', that they don't really spin, and probably don't even have an intrinsic mass!

Why are people so enamored of such counterintuitive, illogical, unphysical, and irrational ideas?

I believe it is a mark of modern sophistication; as if due to our (solely) physical and technological advances, we are now somehow smarter or wiser than Isaac Newton or Immanuel Kant!

As if.

The actual claim seems to be that the math says so. However, we now demand that the math be in accord with the world, rather than the world with the math (i.e. kludged up Lagrangians, infinities, singularities, divergences, ever multiplying forces and fields, etc.)

The world is mathematical, but the world is *not* mathematics.

We shall now try to construct a more sensible phenomenological model of the fundamental particles; their world, and ours.

In the universal model, half integral spin is *the* defining, fundamental, and irreducible representation of any massive inertial particle.

In the sinister universe, the manifestation of half integral spin is the very *essence* of mass and inertia.

A particle of half integral spin can always be assigned a definite position in time and space. No other particle may occupy this same position in space at the same time.

The neutrino:

The neutrino is a tiny blob of rotating matter.

The neutrino may be considered to be a point particle, just as the earth, a galaxy, or even a cluster of galaxies may be considered point particles.

We assume the neutrino has no (particle) substructure, but this does *not* mean that the neutrino does not have extension. Who knows?

The electron neutrino *is* a tiny blob of mass of spin ½, spinning to the left.
The electron neutrino has a magnetic moment proportional to its mass

The electron antineutrino *is* a tiny blob of mass of spin ½, spinning to the right.
The electron antineutrino has a magnetic moment proportional to its mass
(and equal and opposite to the magnetic moment of the electron neutrino).

Two electron neutrinos moving under mutual gravitational attraction *cannot* collide.

Two electron neutrinos *cannot* occupy the same point at the same time.

Two electron neutrinos will ultimately repel one another due to the interaction of their magnetic moments.

An electron neutrino and an electron antineutrino, moving under mutual gravitational attraction, *will* collide, and annihilate one another.

We picture this annihilation as the cancellation or interference of the spins of the two particles, yielding a massive, yet *inertialess* particle, with spin 0; the 'virtual photon'.

The photon:

The photon serves at the pleasure of the leptons; either as the virtual gauge boson mediating their interactions, or as a real particle; for the shedding of energy and momentum of any lepton undergoing and/or *resisting* acceleration.

In our example of neutrino antineutrino annihilation, the subsequent virtual photon may couple to any lepton antilepton pair, *or*, to two real photons. The photon is a self-coupling gauge boson, because it carries gravitational charge (mass-energy).

A bound electron can only make an energy transition (e.g. in the hydrogen atom) if the subsequent change in angular momentum of the system is +/- 1. This is because the electron spin has to flip to produce a real photon of spin=1 as it *accelerates*.

The photon is produced (generated, released) by the electron. It does not pop out of the vacuum. It is not an excitation of a field.

The photon is energy, momentum, and spin 'shed' by the electron

The photon carries just enough quantities to perform its job, which is transferring energy and momentum (and spin) between interacting fermions!

We know the energy of a photon is proportional to its frequency. However, when a photon loses or gains energy, it does not change in 'size', and its spin remains a constant.

The only physical attribute the photon possesses is spin. This spin can be positive or negative relative to the transverse direction of the photon.

We hypothesize that the frequency, nu, of a free photon is the rate at which the photon spin flips, changing polarization harmonically as it barrels toward its target at the speed of light. This 'rate of spin flip' would be what would give the photon a particular energy and momentum.

Our new model of the world is a mechanical model; albeit without the gears, wheels, and pulleys, that Maxwell and Faraday imagined.

Fundamental particles are spinning mass-energy. Half integral spin is the origin of inertial mass. The mass-energy of a particle consists entirely of rotational kinetic energy. This rotational kinetic energy is due to the particle's spin, *and* spin precession about the direction of the linear momentum of the particle.

The linear momentum of a particle (or more specifically, impulse) is due to a particle's helicity flipping harmonically at the characteristic frequency that defines the energy and 'wavelength' of the particle.

This is true for the photon, as we have argued just above, as well as for the electron, as we shall demonstrate in the next section.

Particle in a box:

Let's begin with the classic example of the particle in a one dimensional box.

A particle (e.g. an electron) in a box bounces back and forth between the walls. The particle follows a well defined path and physically traverses *all points* lying between the walls of the box.

The current interpretation of quantum mechanics posits there are points where the electron never is; points the electron does not pass through; points along the trajectory where the electron essentially ceases to exist!

Of course, these points are the nodes where the quantum mechanical wave function of the particle (a sine wave) passes through zero.

In our new model, these nodes represent points where the electron is actually *physically unable* to interact because it is undergoing a 'spin flip', or a 'change in helicity'.

We now understand the spinor nature of the electron, and why it's spin is 'double valued'.

As the electron travels along, its spin precesses once about the direction of travel (it's 'proper' or measured helicity, say) and then the spin flips, going around once the other way; ad infinitum.

At the instant of the helicity flip, the electron is not able to interact with 'real' photons.. These points correspond to the nodes of the quantum mechanical wave function.

We can now interpret the probabilities generated from squaring the wave function to indicate where a particle is actually able to physically interact and *be measured*, rather than where the particle is.

In addition, like a swimmer doing laps in a pool, the boundary conditions of our box dictate that the electron must perform a helicity flip when it bounces off the walls; hence the wave function must yield a zero expectation value at the boundaries of the box.

This also explains the nature of the Bohr orbits in the hydrogen atom, and why they must correspond to integral values of the electron wavelength.

In our new model, the electron is an inertial (i.e. spin ½) blob of mass-energy/charge, spinning to the left. The axis of the electron spin and the principal axis of rotational inertia of the electron are aligned or 'projected' along the direction of motion of the electron and precess about this direction with a frequency proportional to the total mass-energy of the electron.

This precession frequency corresponds to the de Broglie wavelength of the electron;

$$v = E/h == m/h \tag{66}$$

$$\lambda = h/p = h/mv \tag{67}$$

with wavenumber

$$k = 2\pi/\lambda \tag{68}$$

and so it takes 2π radians for electron spin to precess once about the direction of motion.

However, since the electron is a 'spinor', it takes 4π radians, or two revolutions of the spin vector, for the precessing *angular momentum vector* to return to its original value and helicity.

A free electron of fixed helicity, executes a 'polarization flip' every 2π radians, performing a 'complete revolution' every 4π radians. This spin flipping is what gives a particle its oomph!

Particle in a box:

```
                                                              ^
                                                               \
                                                                \
                        |------------------<-----------------------\|
                                                            ^
                                                             \
                                                              \
                        |------------------<-----------------------\---|


                        |------------------<-----------------------/---------|
                                                                 /
                                                                /
                                                               /

   node                 |----------<----------------.-----------------------|

                                                      ^
                                                     /
                                                    /
                        |----------<-------/------------------------|

                                              ^
                                             /
                                            /
                        |-----<---/--------------------------------|



                        |---------\-------------------------->------------|
                                   \
                                    \
                                     \

   artist's conception
```

To reiterate, an electron in a box bounces back and forth between the walls. The electron follows a well defined path and physically traverses *all points* lying between the walls of the box. The behavior of the electron spin vector is illustrated on the previous page.

The wave function for an electron in a box, indicates where, when, and how 'efficiently', the electron is able to interact.

Figure 2 shows the probability distribution for the state n = 2.

The electron spin precesses around the direction of motion, performing a helicity flip every 2 pi radians, and completing one complete 'revolution' every 4 pi radians.

At the point x = L/2, the electron is undergoing a flip in helicity, where its spin is effectively zero, and hence it cannot interact with our experimental probe (e.g. a photon).

Similarly, at the walls of the box, the electron must perform a helicity flip as it bounces, exchanging a virtual photon with an electron in the wall of the box.

At the point x = L/4, the electron helicity is in 'full bloom' and it can interact with a real photon.

The electron oscillates harmonically between the ability to engage in real and 'virtual' interactions

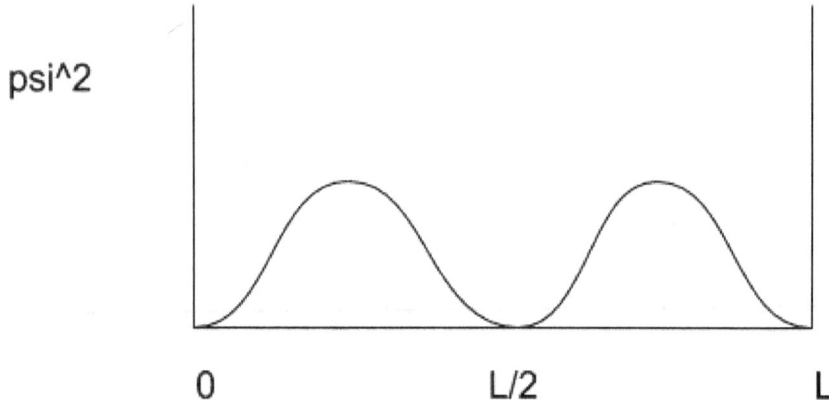

Figure 2: Probability distribution for a particle in a box; n = 2.

Hidden Variables:

We have discovered the hidden variables of quantum mechanics.

They are called spin, mass, and charge!

More specifically, the hidden variables are; the *phase* of the particle wavefunction, the spinor nature of the electron, and the double valued state functions of the electron.

In the standard model, the phase of the wave function, and the spinor nature of the electron, are considered curiosities and dismissed, since they do not contribute to the expectation value of any observable when one squares the wave function.

In our model, the phase of the particle wave function is *real,* and has real, observable, and measurable effects.

In our model, subatomic particle interactions are completely deterministic, and there exist many physical and 'timing' constraints on these interactions, since the electron, as a spinor, can only interact at certain times, and in certain ways; alternatingly interacting with real and virtual photons, as its helicity oscillates back and forth, as it travels a *well defined* path.

We speak of interactions as characterized by the exchange of real or 'virtual' photons.

The irony is that the interactions that occur via 'virtual' photon exchange may actually be considered the 'more real'.

Real photons emitted by a particle may interact with a subsequent particle, but this is completely accidental. Only human beings employ real photons to influence other particles.

For this reason, we prefer to characterize interactions as extrinsic vs. intrinsic, external vs. internal, or accidental vs. incidental.

They are all real interactions.

The single slit experiment:

In our model, the characteristic 'interference' pattern observed on the screen or photographic plate in a single slit diffraction experiment, is not due to two separate electron waves arriving at the screen in or out of phase and reinforcing or canceling each other out.

Instead, *each separate electron* arrives at the screen 'in or out of phase' for being able to interact with the photographic plate.

An individual electron can only darken the photographic plate if it arrives with the proper phase. Just like our particle in a box, the oscillating helicity of the electron has to be in the process of 'flipping polarization' when it arrives at the screen (i.e. the electron must have helicity ~zero to interact with the surface of the plate).

Electrons that do not successfully interact with the surface of the plate will carry on until they *do* find something to interact with.

We suggest a single slit experiment where the photographic plate is replaced with several layers of silicon, analogous to tracking detectors employed in high energy scattering experiments.

Our prediction is, of course, that the electrons which fail to contribute to the pretty pattern on the top layer, will show up in subsequent layers; perhaps displaying secondary pretty patterns.

The double slit experiment

We can employ similar arguments to explain double slit interference patterns, although in reality the situation is much more complicated.

In the single slit experiment, it does not really matter what happens before the slit. The slit is essentially the source of the electrons and the electrons have a characteristic spread determined by the width of the slit and the uncertainty principle.

For the double slit experiment, the environment of the electrons before the two slits *does* matter.

Although an electron can only pass through one slit or the other, it feels the 'potential' of both slits.

Remember, our 'free' electrons are actually exchanging virtual photons with *everything*. On their approach to the two slits, they are interacting with the wall housing the two slits, and they are interacting with the screen (which is their final destination) by virtual photon exchange via the two slits.

How this could actually be cast and analyzed in terms of potentials is unclear!

We suggest a similar double slit experiment, where the the two slits are replaced by two similarly sized and spaced metal plates, set to a negatiive retarding potential; thus, scattering the incoming electrons back toward the source and an appropriately placed screen.

The wave function:

The wave function is longer mysterious or spooky.

You can no more predict the position of an electron bouncing around in a box than you can predict the position of an ideal gas molecule bouncing around in a box.

The wave function is a mathematical function invented by human beings. It has nothing to do with the particle 'itself'. The wave function is *not* the particle and does not 'collapse' when one makes a measurement anymore than a momentum vector describing the particle would collapse.

The wave function is a probabilistic determination of what energy, momentum, and position values we can expect to measure for a particle *we cannot see*.

The wave function is not a real, physical thing. It has nothing whatsoever to do with the particle; either physically, or 'metaphysically'.

The wave function does collapse.

It is a completely meaningless notion and a *non problem*.

Quantization:

Ultimately, quantization boils down to the satisfaction of boundary conditions.

Virtual photons can only attach to/interact with an electron when the electron has an effective helicity of zero (during its polarization flip) because virtual photons are spin zero.

This fixes the 'frequency' of the virtual photon coupling two interacting electrons, since virtual interactions, or the exchange of energy and momentum, can only occur when both electrons have helicity $\sim= 0$.

When an electron reaches 'peak helicity', it is able to interact with (i.e. absorb and emit) real photons the most readily and the most efficiently.

Magnetic moments:

Previously, we introduced the universal model generalization for the formula of the electron magnetic moment;

$$\mu_e = (e/m_e)(m)(\hbar/2m_e) \tag{69}$$

This formula can be interpreted as

$$\mu_e = (\text{coupling constant})*(\text{mass})*(\text{angular momentum per unit mass})$$

So, the electron is one unit of angular momentum per unit mass per unit volume of space; or one unit of inertial, half integral, angular momentum per unit mass.

$$e \Rightarrow h/4\pi m_e == (\hbar/2)/m_e == L/m_e \tag{70}$$

Inserting the formula for the relativistic mass into equation (5) we get

$$\mu_e = (e*\hbar/2m_e)(1/(1 - v^2/c^2)^{1/2}) \tag{71}$$

We then make the usual series expansion to obtain

$$\mu_e = (e*\hbar/2m_e)(1 + \tfrac{1}{2} v^2/c^2 + \tfrac{3}{8} v^4 c^4 + ...) \tag{72}$$

and find the magnetic moment of the electron increases with velocity, *as expected*.

The electron is essentially the vector L/m_e. This vector precesses about the axis of the direction of motion with a frequency; $\nu = E/h = m/h$. As the speed of the electron increases, the frequency of precession increases, and the mass of the electron increases.

The surprising 'spinor' nature of the electron is due to the angular momentum vector flipping helicity/polarization every 2π radians.

Even though the mass and magnetic moment of the electron increase with the velocity, the electron angular momentum is always $\hbar/2$.

We finally know what has been waving all this time!

The waving of the wave equation/wave function represents the periodic precession of the electron spin about the direction of travel of the electron!

N.B. The wave equation describes *many* physical phenomena that don't really wave.

Spinoring:

We keep saying the "helicity or polarization" of the electron 'flips' every 2π radians.

Technically, of course, this terminology is incorrect, because even though there *is* 'flipping' going on, the helicity of the electron never changes!

Instead, we shall say the electron is 'spinor-ing', as illustrated in Figure 3.

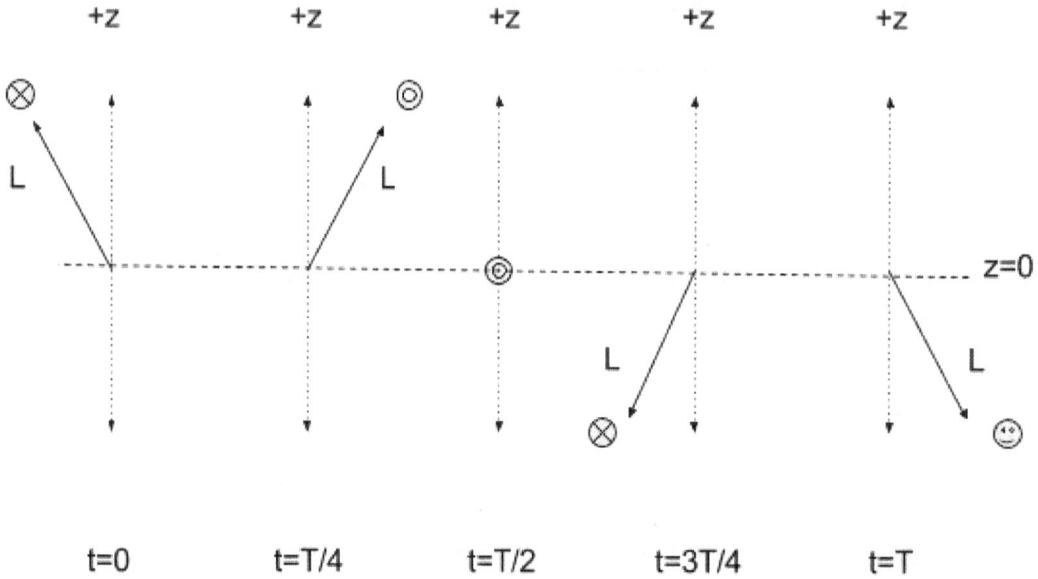

Figure 3: At rest with an electron traveling the the z-direction. The spin angular momentum vector 'precesses' about the direction of motion, tracing out a closed, three dimensional figure eight (a string!). The x symbol represents motion into the page. The dot symbol represents motion out of the page. At time T/2, we see the the angular momentum is *perpendicular* to the direction of travel. (This is when the electron engages in 'virtual' interactions.)

We also can see, that although the angular momentum vector is 'spinor-ing', the polarization, or helicity, of the electron does *not* change, and is constant!

The neutrino magnetic moment is

$$\mu_\nu = (m)(\hbar/2m_\nu) \qquad (73)$$

Inserting the formula for the relativistic mass of the neutrino into equation (73) we see

$$\mu_\nu = (\hbar/2)(1 + \tfrac{1}{2} v^2/c^2 + \tfrac{3}{8} v^4 c^4 + ...) \qquad (74)$$

And, we must conclude that *the neutrino is one inertial quantum of action* as we proposed in "On Matter, Mass, and Motion". We can represent the neutrino symbolically, as we did for the electron in equation (70), and write

$$\nu_e \Rightarrow \hbar/2 == L \qquad (75)$$

Actually, equation (74) shows that *there is only one neutrino*. The three neutrinos differ only by their **velocity.**

We can now see that neutrino mixing is due to elastic collisions of the neutrino with matter, decelerating the neutrino, and changing its flavor. </blink>

The tau lepton is the most massive of the three charged leptons, so when it 'decays', it emits the highest velocity, and hence, most massive of the the "three" neutrinos, etc.

The tau is shedding energy and spin, because what else is there?

We can also derive the magnetic moment of the neutrino using the hand waving arguments usually assumed for the electron magnetic moment.

$$\mu = (\text{mass current})(\text{area}) = (m/t)(A) = m(L/2m) = \hbar/2 \qquad (76)$$

Finally, we reach way back to Physics 101 and recall the formula for angular momentum

$$L = mvr \qquad (77)$$

The angular momentum of the neutrino is h and we assume it spins with angular velocity c.

$$h = (m_\nu)(r_\nu)c \qquad (78)$$

Now, we can solve for the *Compton radius of the neutrino*; the smallest probable distance.

$$r_\nu = h/(m_\nu)*c \qquad (79)$$

The photon, neutrino, and electron:

The photon is essentially a perpetual motion machine! The photon polarization oscillates harmonically at a frequency proportional to its energy, nu = E/h. This is an unorthodox picture of the photon polarization, but we note it satisfies the orthogonality condition on the photon wave vectors; $\mathbf{k} \cdot \varepsilon = 0$. The photon is corpuscular! (Newton: 2, Everybody else : 0)

The photon is one 'free', massive, but inertialess, unit of angular momentum; L = h.

The neutrino is one 'bound', massive, unit of angular momentum *per unit space*; L = hbar/2.

The electron is one 'bound' unit of angular momentum per unit space *per unit mass*; L = hbar/2.

Particle interactions:

Take a look at the angular momentum vector of the electron at time, t=0, in Figure 1. Here, we say the electron helicity is in 'full bloom' and it is able to absorb a real photon (*if* its polarization is also 'blooming'), increasing the rate of precession of the electron angular momentum vector, and hence the electron mass.

Consider the case of partial transmission and partial reflection of light from a thin sheet of glass. Photons in 'full bloom' will be transmitted. Photons with polarization ~0 will be reflected.

Similar arguments can be made for electron tunneling. If an electron arrives at a potential barrier 'out of phase for reflection', *and,* the electron 'wavelength' is comparable to the 'height' of the potential barrier, the electron will 'tunnel' through!

Quantum field theory:

In our model, particles are not created and destroyed. Instead, particles absorb, emit, merge with, and *shed* one another.

For example, an electron and a positron do not 'annihilate', producing a virtual photon.

The electron and positron have equal and opposite spins, ½. They merge to become a photon of spin 0. The photon then splits into the particle antiparticle pair demanded by the situation.

In our model, particle interactions are a continuous flux and flow of energy and momentum, *flowing only one way;* futureward. Particles interact by exchanging units of *angular momentum*.

Particle families:

In our new model, we have replaced the quarks with leptons. However, there is not a one-to-one correspondence between the two species, and 'the mapping' will not necessarily be the same between the different particle groups of the standard model.

In our model, the first generation of particle families is now imagined as follows:

proton = (u,u,d) ⇒ (e+,e-,e+)

neutron = (u,d,d) ⇒ (e+,e-,$\bar{v_e}$)

electron ⇒ electron

electron neutrino ⇒ electron neutrino

In the standard model, the second generation consists of the strange quark, the charm quark, the muon and the muon neutrino.

The charm quark and anti-charm quark can form temporary bound states in high energy collisions ("charmonium" a.k.a the J/Psi meson) which then decay into a muon antimuon pair or an electron positron pair. The strange quark has yet to be seen in such a bound state. Hence, for the second generation, we make the following substitutions;

charm quark ⇒ muon

strange quark ⇒ muon antineutrino

Similarly, for the third generation, we have:

bottom quark ⇒ tau meson

top quark ⇒ tau antineutrino

We still assume the relationships concerning mass, charge and spin derived for the electron and the electron neutrino hold for the higher generations as well.

Why is there more than one particle family?

What dictates the energy differences between the three particle families?

We hypothesize that when generating high energy leptons, at some point it becomes more expedient for 'Nature' to generate a muon rather than an electron with a velocity v/c ~= 1.

This appears evident from equation (74) and our new theory of the neutrino.

If true, where is this point? Is there a hard and fast rule governing when to chose the higher family particle, or is there some probabilistic indeterminacy involved?

In our model, the muon 'sheds' energy and spin in the form of its neutrino. The energy and spin shed ensures that the resulting virtual lepton propagator has spin = 0, and is *massless*.

This requirement places constraints on the energy and momentum of the initial and final states in muon (and tau) decays, and should help to explain the choice of the final state lepton from the three particle families in a particular decay, as well as some of the current mysteries surrounding lepton universality.

We can solve for this threshold, by assuming the muon rest mass is equivalent to the largest *allowable* relativistic mass of the electron.

$$m\_mu = m\_e/(1 - v^2/c^2)^{1/2} \tag{79}$$

$$(v^2/c^2)_{THRESHOLD} = 1 - (m\_e/m\_mu)^2 \tag{80}$$

A similar calculation will result in the velocity threshold between the mass of the muon and the mass of the tau.

Group theory:

Remarkably (and necessarily, perhaps), the universal model retains the SU(3)SU(2)U(1) group structure of the standard model.

However, these groups do not generate 'exchange quanta'. In addition, the color charge is no longer the basis for the group SU(3). Instead, we propose the three charged leptons; the electron, the muon, and the tau.

The group SU(3) reflects an exact symmetry of the universal model, operating on both the charged lepton basis

e = (1,0,0) ; muon = (0,1,0) ; tau = (0,0,1)

and the neutrino basis

nu_e = (1,0,0) ; nu_muon = (0,1,0) ; nu_tau = (0,0,1)

The neutrinos have the same mass hierarchy (~1,100,1000) as the charged leptons.

The eigenvalues, q, of the charge operator are degenerate for the charged lepton basis, but the eigenvalues q/m are non-degenerate, as are the masses themselves.

We can use the step up and step down operators (I+ and I-), composed of the Lambda_i matrices of the standard model, to transform between the particles in each of the two independent bases. Neutrino mixing, anyone?

We can also borrow the formalism of flavor SU(3) by making the substitutions

up → electron

down → muon

strange → tau

and generate all the pseudoscalar mesons, for example;

pi^0 = 1/(2)^½ (e e^bar - mu mu^bar)

Finally, we can generate particles by mixing the charged lepton and neutral lepton bases, and the neutral lepton bases, using the Lambda_i.

The fundamental lepton:

We call our fundamental lepton, the tetrahedron, in honor of Plato; the Man, the Myth, and the Legend!

Our tetrahedron may, or may not, have four sides; but we will assume it has three principle axes of rotational inertia.

In our theory, *all inertia* is rotational inertia. Trying to accelerate an elementary particle is analogous to trying to walk around a high school physics lab while holding a spinning bicycle tire.

In addition, the mass-energy of a particle is due to the rotational kinetic energy about its axis of rotation (at rest), as well as an additional rotational kinetic energy which arises due to the the rate of spin precession of the rest mass around the direction of travel for a particle in motion.

We imagine the three principle axes of inertia of the neutral tetrahedron to correspond to the three neutrino masses, and the three principle axes of the charged tetrahedron to the masses of the electron, the muon, and the tau.

More specifically, in light of equation (70), there is one charged tetrahedron with three principle axes of rotational inertia (L/m_e, L/m_{mu}, L/m_{tau} ; $L = \hbar/2$) and one neutrino to serve them all!

There are now three electromagnetic coupling constants; e/m_e, e/m_{mu}, e/m_{tau}.

The step up and step down operators composed of the Gell-Mann matrices of SU(3), are actually geometrical rotations from one principle axis of inertia to another. When the tetrahedron flips from one axis to another, it sheds its energy and momentum in the form of its corresponding neutrino.

There are three particle families because there are three dimensions of space.

Similar rotations in "weak isospin" space transform the neutral tetrahedron into its charged counterpart.

What is this crazy isospin space? I don't know!

We *do* know that by using various combinations of the step up and step down operators of SU(2) and SU(3), plus the parity operator, we can write any quantum mechanical interaction current in terms of the 'fundamental' neutrino neutral current.

Weak isospin:

Geometric rotations of the three principle axes of inertia of the tetrahedron about the axis of *spin*, generate the three lepton families, transforming one into another; all manifestations of a single particle.

These rotations correspond to the transformation matrices of the group SU(3) of the standard model.

Weak isospin rotations transform the neutral tetrahedron into the charged tetrahedron.

It seems these do not correspond to any real, physical, rotation.

We speak metaphorically of the electron as an excited neutrino; its electric charge a consequence of quantum mechanical electromagnetic induction (qemf).

However, qemf is quantized, so the electron cannot decay or collapse into a neutrino.

Charged leptons can 'decay' into one another, but the charged tetrahedron is conserved.

In conclusion, there are two fundamental leptons. One is neutral and one is electrically charged.

The leptonic table:

LEPTONS   ANTI-LEPTONS

electron	electron neutrino	PARITY ⇔	electron antineutrino	positron
⇐	CHARGE	MASS ↕	CHARGE	⇒
muon	muon neutrino	PARITY ⇔	muon antineutrino	anti-muon
⇐	CHARGE	MASS ↕	CHARGE	⇒
tau	tau neutrino	PARITY ⇔	tau antineutrino	anti-tau
⇔	weak isospin	mass isospin ↕	weak isospin	⇔

TABLE 1: The leptons and their interrelations.

Any lepton can be 'generated' from any other by the appropriate applications of the parity operator, the weak isospin operator, and our newly proposed 'mass isospin' operator.

The Heisenberg uncertainty principle:

The Heisenberg uncertainty principle reflects the fact that our experimental probes perturb or destroy the microscopic system under investigation.

It is *not* a ghostly feature of reality reflecting the murkiness of the physical properties of the particles and/or their *very* existence

The Pauli exclusion principle:

Now that our most fundamental lepton, the electron neutrino, has a magnetic moment,

and *we now have a completely mechanical model of subatomic particle interactions*,

we can understand (as most people have long suspected) that the Pauli exclusion principle is due solely to the electromagnetic interaction between leptons -- which act like tiny little refrigerator magnets!

There are no more mysterious quantum mechanisms.

The mechanical universe:

Bicycle wheels and refrigerator magnets!

The hydrogen atom:

One of our conclusions from investigating the boundary conditions of an electron in a box was that analogous boundary conditions must apply to the electron in the hydrogen atom

We are bringing back closed Bohr orbits which correspond to integral values of the electron wavelength!

In addition, we assume the first Bohr orbit is *circular,* even though the angular momentum of the system has a value of zero.

N.B. We made this same assumption in our analysis of the neutrinium system which then allowed us to compare our results to those of general relativity.

In the ground state of the hydrogen atom, the spins of the proton and electron are aligned (+½, +½) for an angular momentum of +1. The orbit of the electron has an angular momentum of -1. The total angular momentum of the ground state is zero.

The magnetic moments of the proton and the electron are anti-aligned and repel one another keeping the electron from spiraling into the proton.

The stability of the hydrogen atom is now due to the physical necessity of closed electron orbits and the interaction of the magnetic moments of the proton and the electron.

Quantized orbits are now the result of ordinary physical interactions

O.K. -- *now,* there are no more mysterious quantum mechanisms!

The two pillars of twentieth century physics:

The two great pillars of twentieth century physics, general relativity and quantum mechanics, are incompatible and irreconcilable because they are both incorrect.

Conclusion:

Nature is mechanical and *dynamical*. Particles and their interactions can be completely described and explained in terms of fundamental, 'solid', and *real* units of matter, constantly in motion, and continually exchanging energy and momentum.

No more ghosties!

Isaac Newton --   best physicist ever!

Immanuel Kant -- most boring genius!

A quantum mechanical theory of everything:

photons, electrons, neutrinos; interacting synchronously

Books by Greg Feild:

the pentateuch

1. "A quantum mechanical theory of gravitational interactions"
   CreateSpace Independent Publishing, 8/29/2016

2. "Observations on the quantum mechanical nature of gravity"
   CreateSpace Independent Publishing, 10/8/2016

3. "On gravitation and electric charge"
   CreateSpace Independent Publishing, 10/29/2016

4. "On spin, mass, and charge"
   CreateSpace Independent Publishing, 11/29/2016

5. "On angular momentum, acceleration, and absolute motion"
   CreateSpace Independent Publishing, 1/1/2017

the exegeses

6. "The Sinister Universe"
   CreateSpace Independent Publishing, 3/1/2017

7. "On Parity and Isospin"
   CreateSpace Independent Publishing, 4/11/2017

8. "Reflections on the Sinister Universe"
   CreateSpace Independent Publishing, 5/12/2017

the hermeneutics

9. "On Current Physics"
   CreateSpace Independent Publishing, 6/11/2017

10. "A Critical Examination of Classical and Quantum Mechanical Waves"
    CreateSpace Independent Publishing, 6/18/2017

the gospels   :)

11. "On wave particle duality and the quantum of action"
    CreateSpace Independent Publishing, 7/6/2017

12. "On matter, mass, and motion"
    CreateSpace Independent Publishing, 9/14/2017

13. "On action and reaction"
    CreateSpace Independent Publishing, 9/24/2017

14. "A quantum mechanical theory of everything"
    CreateSpace Independent Publishing, 11/5/2017

the compilations

"The Universal Model of Our Sinister Universe: The First Ten Books"
CreateSpace Independent Publishing, 7/2/2017

"The Canons of the Sinister Universe:
The Last Four Books on the Universal Model of Our World"
CreateSpace Independent Publishing, 11/5/2017

Notes: :)

Eternal recurrence:

It's all the same -- day, man ...

-- Janis Joplin

www.ingramcontent.com/pod-product-compliance
Lightning Source LLC
Chambersburg PA
CBHW082206220526
45470CB00010B/3063